Geology and Mineralogy Research Developments

nova
science publishers
www.novapublishers.com

Geology and Mineralogy Research Developments

Gypsum: Sources, Uses and Properties
Matthew N. Ippolito (Editor)
2022. ISBN: 978-1-68507-932-1 (Softcover)
2022. ISBN: 979-8-88697-016-6 (eBook)

Magnetite: Properties, Occurrence and Applications
Manuel J. Wolcott (Editor)
2021. ISBN: N/A (Softcover)
2021. ISBN: N/A (eBook)

A Focus on Mining
Henri Walker (Editor)
2021. ISBN: 978-1-53619-037-3 (Hardcover)
2021. ISBN: 978-1-53619-164-6 (eBook)

Minerals and Their Properties: Novel Approach for Applications
Sanjay J. Dhoble
Anup P. Bhat
Renu Nayar
Bandana Samant (Editors)
2020. ISBN: 978-1-53618-889-9 (Hardcover)
2020. ISBN: 978-1-53618-990-2 (eBook)

A Closer Look at Conflict Minerals
Matias Russo (Editor)
2020. ISBN: 978-1-53618-947-6 (Hardcover)
2020. ISBN: 978-1-53618-966-7 (eBook)

More information about this series can be found at
https://novapublishers.com/product-category/series/geology-and-mineralogy-research-developments/

Joseph E. Glenn
Editor

Geopolymers

Structure, Properties and Applications

nova
science publishers
www.novapublishers.com

Library of Congress Cataloging-in-Publication Data

ISBN: 979-8-88697-197-2

Published by Nova Science Publishers, Inc. † New York

Contents

Preface

Geopolymers are networks of mineral molecules linked with covalent bonds. Geopolymers have a number of commercial applications and are used in medicinal applications, high-temperature ceramics, and more. This volume includes four chapters discussing various aspects of geopolymers. Chapter One reviews some of the modifications, applications, and challenges of waste-derived geopolymer catalysts. Chapter Two investigates the adsorption of Cu(II), Mn(II), Pb(II), and Zn(II) ions to geopolymers obtained from biomass combustion ash. Chapter Three presents a summary of the literature on geopolymer concrete's foundations and disadvantages compared to portland cement concrete, mix proportions that give the best strength, management of fiber additives, performance analysis against burst rate, and resistance to attack.

Chapter 1 - The generation and piling of solid wastes in the industrial, agricultural, and household sectors are worrisome. These wastes often contain hazardous components which can leach into the surface and groundwater and pose long-term threats to human health and the environment. To reduce the piling and motivate resource recovery, they must be reused economically and in aneco-friendly manner. Geopolymers, which are amorphous analogues to zeolites, can be synthesized from various wastes and have emerged as the most promising porous materials having extensive applications. Geopolymeric adsorbents have gained vast popularity in the past, and geopolymer applications have recently extended to the field of heterogenous catalysis. This review discusses the geopolymer modification methods (emulsion templating, direct foaming, demetallation, impregnation, composite technique, etc.) and their catalytic applications to attain a clean and green environment. The catalytic applications discussed in this review focus on environmental governance, with environmental remediation and clean/green production as the main center of discussion. With the summary and critical analysis of available literature, this review is expected to help and encourage the scientific community to find innovative ideas for developing geopolymeric catalysts and their widespread applications in areas requiring intervention.

Chapter 2 - Several industries produce wastewater containing a wide range of pollutants. This includes organic impurities such as dyes, petroleum products, pharmaceuticals, and inorganic pollutants, which mainly contain heavy metals. It is widely believed that both dyes and heavy metals are among the most dangerous contaminants of water systems due to their mutagenic, carcinogenic, and teratogenic effects. Efficient wastewater treatment using conventional methods is often limited due to the low selectivity and high costs of such methods. An ecological, economical, and efficient alternative to conventional methods is the adsorption process on the geopolymers obtained from the by-products of combustion, i. e. the ash from biomass combustion.

The aim of the work was to investigate the adsorption of Cu(II), Mn(II), Pb(II), and Zn(II) ions to geopolymers obtained from biomass combustion ash. The influence of the starting metal concentration in the aqueous solution and the adsorption duration on the adsorption efficiency of the ions studied was discussed. The Langmuir and Freundlich models were used to describe the adsorption equilibrium of Cu(II), Mn(II), Pb(II), and Zn(II) ions from aqueous solutions on geopolymers obtained from the ash of biomass combustion.

Analysis of the results showed that the adsorption efficiency of the tested ions increases over time in the following ranges: Cu(II) 50–95%, Mn(II) 41–92%, Pb(II) 33–89%, Zn(II) 58–90%. Experimental sorption content values are 133 mg/g for Cu(II), 105 mg/g for Mn(II), 348 mg/g for Pb(II), and 132 mg/g for Zn(II). The adsorption of the investigated ions to geopolymers extracted from the ash of biomass combustion is carried out according to a pseudo-second order model. Analysis of the FTIR spectra of geopolymers after adsorption of the tested metal ions indicates a proportion of functional groups, e.g., O-H, N-H, and C-O, by adsorption. Comparison of all tested geopolymers that have adsorbed Cu(II), Mn(II), Pb(II), and Zn(II) ions showed that they have the smallest size for the geopolymer after adsorption of Pb(II) ions and the largest for the geopolymer after adsorption of Zn(II) ions.

In summary, geopolymers from biomass combustion by adsorption of Cu(II), Mn(II), Pb(II), and Zn(II) ions from aqueous solutions can be used. The geopolymers used have good sorption properties compared to the heavy metal ions studied, indicating that they can be used as adsorption agents for industrial and environmental pollutants. In addition, the proposed application is an example of the use of geopolymers as recycled products of combustion by-products.

Chapter 3 - In recent years, different types of special concretes have been produced and used according to the usage principles of the buildings. One of these concretes is alkali active concretes produced without the need for cement production which is known as geopolymer concrete. Their high early strength, less drying-shrinkage, high fire, abrasion & acid resistance, less CO2 emissions, least setting time, and high freeze-thaw & corrosion resistance are reasons this type of concrete has emerged as an alternative to portland cement concrete. This concrete is formed by the reaction of material rich in alumina and silica with the activator. Since geopolymer concrete is more environmentally friendly than cementitious concrete, it has been the subject of many studies. Geopolymer concretes have some disadvantages despite all these advantages. One of them is brittleness. It is known that adding fibers to geopolymer concretes positively affects brittleness and ductility, compressive strength, and flexural strength. Also, mineral additives strengthen the durability of concrete. This work presents a summary of the literature on geopolymer concrete's foundations and disadvantages compared to portland cement concrete, mix proportions that give the best strength, management of fiber additives, performance analysis against burst rate, and resistance to attack.

Chapter 4 - A sulfated geopolymer catalyst was prepared via a three-staged treatment comprising emulsion templating, ion exchange and sulfation treatment. The modification method improved the catalytic performance, surface morphology and porosity. Emulsion templating aided in creating macro-and mesoporosity in the sample, while ion exchange and sulfation treatments enhanced the acidity of the catalyst. The catalysts obtained at each stage of the treatment were characterized by SEM, TEM, XRD, XRF, BET, NH3-TPD, and FTIR analyses. The catalysts prepared at each stage of the sequential treatment were subjected to a dibenzothiophene oxidation reaction, a model reaction for studying the applicability of the sulfated geopolymer catalyst in oxidative desulfurization. The efficiency of the catalyst was enhanced after each stage, and the conversion of dibenzothiophene was increased from 39 to 74%. The catalyst was thus deemed suitable for the preliminary treatment of petroleum fuel for reduction of sulfur content.

Chapter 1

Waste-Derived Geopolymer Catalysts: Modifications, Applications and Challenges

Snehlata Kumari, Anjitha V. and Sonali Sengupta*
Department of Chemical Engineering, Indian Institute of Technology Kharagpur, Kharagpur, West Bengal, India

Abstract

The generation and piling of solid wastes in the industrial, agricultural, and household sectors are worrisome. These wastes often contain hazardous components which can leach into the surface and groundwater and pose long-term threats to human health and the environment. To reduce the piling and motivate resource recovery, they must be reused economically and in aneco-friendly manner. Geopolymers, which are amorphous analogues to zeolites, can be synthesized from various wastes and have emerged as the most promising porous materials having extensive applications. Geopolymeric adsorbents have gained vast popularity in the past, and geopolymer applications have recently extended to the field of heterogenous catalysis. This review discusses the geopolymer modification methods (emulsion templating, direct foaming, demetallation, impregnation, composite technique, etc.) and their catalytic applications to attain a clean and green environment. The catalytic applications discussed in this review focus on environmental governance, with environmental remediation and clean/green production as the main center of discussion. With the summary and critical analysis of available literature, this review is expected to help and encourage the scientific community to find innovative ideas for developing geopolymeric catalysts and their widespread applications in areas requiring intervention.

---

* Corresponding Author's Email: sonalis@che.iitkgp.ac.in.

In: Geopolymers
Editor: 979-8-88697-197-2
ISBN: Joseph E. Glenn

Keywords: catalyst, geopolymer, review, waste valorisation

Introduction

Rapid industrialization and ongoing urbanization have led to the generation and accumulation of massive amounts of solid wastes in our surroundings, which are potential environmental pollutants. The RRR (Resource, Recovery, and Reuse) approach is the correct response to control the generation and accumulation of solid wastes but is not rightfully observed in industries and municipalities. Currently, cities worldwide produce approximately 1.3 billion tons of solid waste, which is expected to increase to 2.2 billion tons by 2025 (Hoornweg, Daniel; Bhada-Tata 2012). In addition to these, industrial solid wastes also contribute billions of tons. The improper disposal of these wastes has long-term detrimental effects on human health and the environment. Often, these wastes contain abundant amounts of valuable entities that can act as raw materials for economically viable eco-friendly products. In search of a sustainable future, the successful use of large-scale industrial wastes, municipal wastes, and plentiful natural materials has become a subject of significant interest.

For decades, hundreds of millions of tons of different industrial wastes such as blast furnace slag, coal fly ash, red mud, steel slag, fayalite slag, waste glass, silica fume, coal gangue, high magnesium nickel slag, volcanic ash, rice husk ash, sugarcane bagasse ash, and waste incineration bottom ash have accumulated in open fields, and created environmental issues due to particle deposition or leaching of harmful contaminants in soil and water. These materials contain high amounts of aluminates and/or silicates and, in some cases, are rich in other alkaline earth metals such as calcium or magnesium. Thus, they can provide desirable resource recovery, which aids in economic and eco-friendly geopolymer synthesis. No-cost and abundant natural materials such as laterite soil and clays like kaolin, bentonite, halloysite, mullite, etc., also contain valuable resources that can be exploited for sustainable growth. The chemical composition of all these materials also resembles that of aluminosilicates and therefore are reasonably suited for synthesizing geopolymers. Scholars in the past have successfully synthesized geopolymers possessing varied properties with these raw materials.

Geopolymers are inorganic polymers of amorphous aluminosilicates consisting of cross-linked silicate and aluminate units connected through shared oxygen atoms into a 3D structure. The negative charge on the tetrahedral aluminate is balanced by the presence of ion-exchangeable alkali ions that give them similar properties to the zeolites (O'Connor et al. 2010). They are synthesized via alkali activation of aluminosilicate materials through a two-step process consisting of two controlling stages; nucleation or dissolution of the aluminosilicate material and crystalline growth. The dissolution of aluminosilicates is rapid at high pH, and these processing conditions and characteristics of the aluminosilicate source affect the geopolymer properties strongly. Nevertheless, acid activation of aluminosilicates to produce geopolymers is also evidenced in a few pieces of literature (Lin et al. 2021) (B. Zhang, Guo, Deng, et al. 2020) (Sellami, Barre, and Toumi 2019). The synthesized geopolymers can be tailored by the correct mix and processing design to optimize properties and reduce cost for a given application (Duxson et al. 2007).

Traditionally, geopolymers are categorized as mesoporous materials, but their porosity can be tailored via emulsion templating, demetallation, composite technique, etc. They also possess excellent ion-exchange capacity, which can be utilized to link other ions or to amplify the acidic, basic, or photocatalytic properties of the geopolymer (Kumari et al. 2021) (José Ramón Gasca-Tirado 2018) (Alzeer and MacKenzie 2018). Sensibly combining these modification techniques is a way forward to synthesizing geopolymeric materials with glorified properties. Due to their high thermal resistance and mechanical strength, these modified geopolymers can either serve as excellent catalysts themselves or can be used as support to load active metals. Owing to their rapid solidification ability, they can also be converted to granules which are best suited for catalytic column studies.

Waste-Derived Geopolymers

As discussed in the previous section, various natural and industrial wastes contain several value-added components that can be recycled and reused, thus promoting resource recovery. They can be fabricated using a wide range of materials such as blast furnace slag, coal fly ash, red mud, steel slag, waste glass, silica fume, coal gangue, high magnesium nickel slag, volcanic ash, rice husk ash, sugarcane bagasse ash, waste incineration bottom ash, kaolin and bentonite clays, etc. For instance, clay which is an aluminosilicate salt possessing a layered silicon-oxygen tetrahedron and alumina octahedron, can be used as a precursor for geopolymer preparation. Although they are naturally available in abundance, some of them require dehydration at elevated temperatures, which makes them slightly expensive. Some new studies are now directed towards utilizing natural kaolin for the preparation of geopolymer, and the geopolymer is synthesized by alkali activation of natural kaolin (Pelisser et al. 2021). This can be expected to expedite the mass production of geopolymers at reduced production costs. Laterite soil is another abundantly available natural material that can be effectively utilized for the production of geopolymers. It comprises aluminosilicates, iron, and aluminum and has traditionally been used for construction purposes due to its corrosion resistance. The geopolymers prepared by laterite soil have been reported to have high mechanical strength and improvised microstructure due to their distinct molar oxidation ratio (Subaer et al. 2019). Apart from laterite soil and kaolin or bentonite clays, other natural materials such as mullite, halloysite, diatomite, and bauxite may also serve as geopolymer precursors, but their low amorphicity and requirement of high-temperature treatment sometimes hinder their applicability in geopolymer production (Cong and Cheng 2021).

The class of industrial wastes, too, is rich in various value-added components. Fly ash which is produced by thermal power plants in voluminous amounts, comprises aluminum, silicon, and calcium as major components. It serves as the most convenient and ecofriendly resource for producing geopolymers, and the positive impacts on the environment are prominent.

Blast furnace slag, a by-product of ironmaking, when transformed into ground granulated blast furnace slag under appropriate cooling conditions, is amorphous, hard, and has pozzolanic characteristics (Jiang et al. 2020). The Si/Al ratio of this material is 1.71-3.67 and is highly reactive to geopolymerization reaction even at temperatures as low as 0°C, and the geopolymer produced by the slag has been reported to possess long term-strength, high porosity, and sulfate resistant characteristics (Lemougna et al. 2020) (W. Li and Yi 2020).

Red mud is another industrial waste that is produced after the soluble part of bauxite is separated by Bayer's process. Bayer's process utilizes sodium hydroxide for the dissolution process, which leaves the by-product red mud highly alkaline (Nie et al. 2019). This red mud has been utilized as an additive to fly ash geopolymers, which reduces the amount of alkali activator required for optimal geopolymerization reaction to occur. The geopolymer thus produced has been reported to possess comparatively higher strength and durability (Liu, Qin, and Chen 2020) (Yang et al. 2019).

Table 1. Waste materials, their sources and modification methods for geopolymer preparation and their properties

Waste material	Source/Industry	Modifying method/ Modifying agent	Properties	Reference
Kaolin clay	Aluminosilicate source	Alkali activation	• High compressive strength • Thermal insulation • Acoustic insulation	(H. Gao et al. 2020) (Pelisser et al. 2021)
Metakaolin	Aluminosilicate source	Dehydration of kaolin at elevated temperatures	• High compressive strength • High bonding strength	(Duan, Yan, and Luo 2016) (Prud'homme et al. 2011)
Halloysite clay	Aluminosilicate source	Dehydroxylation Alkali activation	• High compressive strength	(B. Zhang, Guo, Yuan, et al. 2020)
Halloysite clay + volcanic glass	Aluminosilicate source	Size reduction Alkali activation	• High mechanical strength • Good metal immobilization capacity	(Barrie et al. 2015)
Illite/Smectite clay	Aluminosilicate source	Dehydroxylation Alkali activation	• High compressive strength for samples calcined at 750°C • Low compressive strength for samples calcined at 650°C	(Buchwald et al. 2009)
Bentonite clay	Aluminosilicate source	Alkali activation	• Greater amorphousness • Porosity • Improved sorption capacity	(Jaimes, Montaño, and González 2020)
Laterite soil	Aluminosilicates, Iron and aluminum	Dehydroxylation Alkali activation	• High mechanical strength	(Subaer et al. 2019)
Fly ash	Aluminum, silicon, calcium	Alkali activation	• Good activity for Friedel–Crafts reaction	(Alzeer and MacKenzie 2018)

Table 1. (Continued)

Waste material	Source/Industry	Modifying method/ Modifying agent	Properties	Reference
Blast furnace slag	Aluminosilicate source	Alkali activation	• High fire resistance • High compressive strength • Long term-strength • High porosity • High sulfate resistance	(Cheng and Chiu 2003) (X. Huang et al. 2015) (Lemougna et al. 2020) (W. Li and Yi 2020)
Linz-Donawitz converter (LD) slag	Calcium oxide, Iron oxide, silica, and alumina	Alkali activation	• Highly porous • High sorption capacity	(Sarkar, Basu, and Samanta 2018)
Red mud	Ferric oxide, sodium hydroxide	Alkali activation	• Comparatively higher strength and durability	(Liu, Qin, and Chen 2020) (Yang et al. 2019)
Rice husk ash (RHA)	Silica	Alkali activation	• High surface area • High strength	(Molaei Raisi, Vaseghi Amiri, and Davoodi 2018)
Calcined Water Treatment Sludge (WTS)+RHA	Silicon oxide, aluminum oxide	Alkali activation	• Lightweight structure	(Nimwinya et al. 2016)
Incinerator bottom ash (IBA)	Silicon oxide, aluminum oxide, metallic aluminum	Alkali activation	• Aerated/lightweight geopolymer • High immobilisation capacity • Low compressive strength • Low density	(Z. Chen et al. 2016)
Coal gangue	Silicon, alumina	Thermal activation of coal gangue Alkali activation to produce geopolymer	• Low compressive strength	(W. Zhang et al. 2020)

Silicon is an essential part of geopolymer, and wastes rich in silicon can also be viable for geopolymer production. Rice husk or RHA, a waste of the rice milling industries, has been utilized as a source of silicon for geopolymer production by various researchers. This not only reduces the price of geopolymer production but also saves the environment from the problems caused by their disposal in landfills. As RHA itself possesses a huge surface area, the geopolymer produced by the silicon derived from rice husk can be expected to demonstrate huge surface areas (Pathak et al. 2020). Bottom ash generated after incineration of municipal waste is laden with various heavy metals and can be a potential source of producing geopolymers with specific properties and specific application areas. This reduces the dependency on purchased heavy metal salts and also prevents the leaching of these metals from landfills. They can be treated as additives during geopolymer production, and the geopolymers prepared with the heavy metal laden bottom ash have been reported to possess higher strength than those produced without the addition of bottom ash (Rutkowska et al. 2018). Apart from that, the waste catalyst residue of various industries can also be used as excellent raw materials for geopolymer production. But their amorphicity is critical from the point of view of geopolymer production. The details of the waste materials, their sources, and modifying agents and modified properties are tabulated in Table 1.

Geopolymer Modification Methods

Broadly, the properties of geopolymers can be tailored via a wide range of modification methods. They can be made into microspheres via the suspension-dispersion-solidification method (Tang et al. 2015) (Tang et al. 2019) (Novais et al. 2019), acidic via ion-exchange with NH_4^+ and sulfate impregnation (Alzeer and MacKenzie 2018) (Alzeer, MacKenzie, and Keyzers 2016) (Alzeer, MacKenzie, and Keyzers 2017) (Kumari et al. 2021), porous via demetallation (Kumari et al. 2021) (Alzeer, MacKenzie, and Keyzers 2017), reactive emulsion technique (Bai et al. 2018), foaming technique (Landi et al. 2013), and can also be modified via incorporation of metal/metal-oxo cationic species or transition metal species or noble metals into the geopolymer matrix by ion-exchange technique (Sazama et al. 2011). In addition to these, they can also be modified via impregnation of metal oxides or alkaline earth metals and doping or loading of transition metals

(Sharma et al. 2015) (Y. J. Zhang, Zhang, Kang, and Zhang 2017) (Y. J. Zhang, He, et al. 2017) (Candamano et al. 2015).

Alkali activation, an important step of geopolymerization reaction, is highly temperature-dependent (Tang et al. 2019). Geopolymer slurry tends to solidify rapidly at elevated temperatures, and the property of rapid solidification can be utilized to prepare inorganic polymer microspheres. Tang et al. (Tang et al. 2019) prepared geopolymer microspheres via the suspension-dispersion-solidification method, wherein geopolymer slurry was injected into a silicone oil bath kept under rapid stirring. As soon as the slurry was injected, it was broken into small droplets by the shear force of the silicone oil, and as the oil was maintained at a high temperature, the droplets were solidified rapidly and settled at the bottom (Tang et al. 2019). When the silicone oil was maintained at low temperatures, the droplets did not solidify but agglomerated at the bottom of the reactor (Tang et al. 2019). However, increasing the temperature beyond 80°C was reported to inhibit the reaction of slag particles, thereby hindering complete geopolymerization. Tang et al. (Tang et al. 2015), in another study, used a PEG-600 medium maintained at 70°C for the preparation of geopolymer microbeads from a foamed geopolymer slurry. The polyethylene glycol medium was stated to promote the condensation rate of geopolymer formation by dissolving the water released during the polycondensation reaction between silicon-oxygen tetrahedral and alumina tetrahedron. Novais et al. (Novais et al. 2019) too prepared geopolymer microbeads in the same way by injecting a foamed geopolymer slurry in a polyethylene glycol medium maintained at 85°C. The utilization of geopolymeric beads in catalytic applications is not common, and till date they have been mainly utilised towards adsorption applications, However, as geopolymer powders cannot be directly used in packed beds, geopolymer microspheres can be utilised in continuous operations.

Geopolymers are cationic exchangers with exchangeable charge balancing cation present in the cavity of their 3D cage-like structure. Their negatively charged structure, which is balanced by alkali cations, such as Na^+ or K^+, can be completely hydrated and mobilized, which gives them an ease of ion exchange when they come in contact with chloride or nitrate solutions of the cation to be ion-exchanged (José Ramón Gasca-Tirado 2018). The maximum number of ion-exchangeable cations is equal to the negative charge of the geopolymer structure and can be determined by the elemental analysis of aluminum (José Ramón Gasca-Tirado 2018). This ability to undergo ion exchange makes geopolymers capable of transforming their H-form by ion exchange with nitrate or chloride salts of ammonium. The alkali metal cation of geopolymer is exchanged with ammonium, which after calcination, is decomposed and transformed to its H-form (Kumari et al. 2021) (Alzeer and MacKenzie 2018). Ion-exchange treatment generates both Lewis and Bronsted acid sites in the geopolymer, and the same can be identified via FTIR spectra of the samples (Alzeer and MacKenzie 2018). The Bronsted acid sites result from H-bonded vicinal silanols and bridging hydroxyls, while the Lewis acid sites result due to extra framework aluminum species generated during thermal treatment (Alzeer and MacKenzie 2018). Apart from ion exchange with ammonium salts, geopolymers can be made acidic by impregnation of sulfate ions too. Kumari et al. (Kumari et al. 2021) reported the preparation of a sulfated geopolymer with enhanced acidity via sequential treatment with desilication, ion exchange, and sulfation treatment, which increased the acidity of the sample from 1.70 to 58.53 mmol/g. The highest increase in the acidity of the samples was evident after sulfation treatment, indicating that the acidity of geopolymeric catalysts can be widely tailored by the reported treatment.

The porosity of geopolymer materials can be tailored by treatment techniques such as demetallation, reactive emulsion templating, composite technique, foaming technique, ice templating, replica method, and additive manufacturing too.

Demetallation is defined by the selective removal of framework atoms (Al, Si) from a geopolymer matrix through acidic or alkali treatment. In zeolitic materials, demetallation can be achieved by acid or alkali treatment and steaming or radiation treatment (Serrano, Escola, and Pizarro 2013). This treatment involves desilication and dealumination of geopolymers and can create mesopores or macropores in the sample. Dealumination or desilication can not only create meso- or macropores in the sample but can also alter the acidity of the samples by changing its Si/Al ratio (Schwieger 2016). This change was evident in the paper by Kumari et al. (Kumari et al. 2021), wherein the acidity of the desilicated sample was enhanced to 8.35 mmol/g from 1.70 mmol/g. Dealumination, i.e., selective removal of aluminum from geopolymer, can also aid in the creation of pores and result in extra-framework aluminum, which are potential Lewis acid sites. Although Lewis acid sites can be advantageous for some reactions, in some cases, they turn out to be disadvantageous by promoting coke formation (Guisnet and Ribeiro 2011). Therefore the choice of modification method primarily depends upon the reaction under consideration. Both desilication and dealumination result in the introduction of porosity in geopolymers; the pores created by dealumination are poorly connected, while those created with desilication are highly interconnected. The enlargement of pore size after sequential treatment with desilication and dealumination combined with ion exchange was evident in the results reported by Alzeer et al. (Alzeer, MacKenzie, and Keyzers 2017).

The reactive emulsion templating technique is another popular method to create macropores in a geopolymer. This method is primarily based upon the saponification reaction between alkali-rich geopolymers and triglycerides contained in oils. The saponification generates soap and glycerol, which can be washed with hot distilled water, butanol, or hexane, forming a macroporous geopolymer. Literature suggests the successful formation of porous geopolymers with canola oil, paraffin, and waste vegetable oil (Sharma et al. 2015) (Medpelli, Seo, and Seo 2014). Pores in size range of 5-40 μm were visible in the canola oil treated samples prepared by Medpelli et al. (Medpelli, Seo, and Seo 2014), while their other samples possessed pores mainly in the mesopore region.

Foaming is a thermodynamically unstable process wherein gas bubbles in the wet foams undergo continuous drainage, Ostwald ripening, and coalescence to minimize the system's overall Gibbs free energy. Wet foams are unstable and generate a large and wide range of pores. Stabilizing agents such as surfactants, particles, or fibers can be used to reduce this. During the direct foaming process, air or gas are generated in a homogeneous slurry via mechanical frothing, microwave frothing, and blowing agents (physical and chemical). The pore structure of the modified geopolymer can be controlled by controlling the amount of blowing agent, surfactant, the viscosity of the slurry, composition, mixing speed, and curing temperature. Physical blowing agents involve phase change, i.e., the gas dissolved in the slurry under high pressure is desorbed by decompression. In contrast, chemical blowing agents involve chemical agents which release gas under a highly alkaline environment. This approach has not yet been utilized to produce hierarchical geopolymers. Chemical blowing agents involved in the production of hierarchical geopolymers include hydrogen releasing metal powders (aluminum or silicon), oxygen releasing agents (hydrogen peroxide, sodium hypochlorite, sodium perborate), and ammonia and sulfur dioxide releasing agents (AlN and $FeSO_3$) containing industrial wastes generated by aluminum waste processing industries (Sanjayan et al. 2015) (Medri et al. 2013) (Bai and Colombo 2017) (Böke et al. 2015) (Bajare, Bumanis, and Korjakins 2014). Stabilizing agents decrease the surface tension of the air/slurry system and stabilize the wet foam by decreasing Ostwald ripening, liquid film rupture, and drainage, thereby facilitating better control of the cell size and size distribution and the ratio of open and closed cells. They influence the morphology and cell structure of the foams and can be non-ionic, cationic, anionic, and proteinic.

Particles and fibers are added to geopolymer slurry as thickening agents to increase the slurry's viscosity and other rheological properties and influence the final pore structure and pore volume. The control of porosity in geopolymers with the help of stabilizing agents such as sodium lauryl sulfate, egg white, rice starch, carbon or cellulose fibers, oleic acid, lime, etc. have been reported in the literature (Esmaily and Nuranian 2012) (Bai and Colombo 2017) (Prud'homme, Joussein, and Rossignol 2015) (František et al. 2014). Apart from these, foams can also be generated without blowing agents with the help of air foam generators, foam generating tank, or special mixing blades (Bai and Colombo 2018). In this case too, the bubble coalesce can be controlled by adding stabilizing agents. The combined saponification/peroxide route is another method developed by Cilla et al. (Cilla, Morelli, and Colombo 2014) which can create a high degree of hierarchical porosity in geopolymer samples. The combined action of frothing created by hydrogen peroxide and surfactants formed by saponification reaction leads to foams possessing a high degree of pore interconnectivity.

Modification of geopolymers via incorporation of transition or noble metal ions through ion exchange has been attempted by Sazama et al. (Sazama et al. 2011), who developed Fe, Cu, Co and Pt incorporated geopolymers for catalytic reduction of nitrogen oxides by ammonia and the total oxidation of volatile organic compounds. At first, they prepared an NH_4^+ form of geopolymer by exchanging the geopolymer with ammonium nitrate solution. They subsequently exchanged the NH_4^+ form of geopolymer with metal salts of Cu and Co, while Fe and Pt geopolymers were prepared by the impregnation method. This step leads to the generation of transition or noble metal incorporated geopolymers with enhanced catalytic properties.

As divalent Cu(II) cationic species exhibit high activity and selectivity in catalytic reduction of ammonia as compared to Cu-zeolite based catalysts, the Cu ion-exchanged geopolymer catalyst was reported to be efficient due to the presence of Cu^{2+} ions at the cationic sites (Kieger et al. 1999) (Sazama et al. 2011). The Co-ion exchanged catalysts prepared by Sazama et al. (Sazama et al. 2011) were employed in the total oxidation of VOCs. The negative charge of the geopolymer network in the case of Co-exchanged geopolymer was balanced by extra network solvated $[Co^{2+}(H_2O)_6]^{2+}$ complexes.

Table 2. Geopolymer modification methods, details of the precursor, activator, modifying agent, and the modified properties

Modification	Precursor	Activator	Modifying agent	Modified properties	Reference
Geopolymerization using alkali activator	Coal fly ash	$KOH + K_2SiO_3$	Oleic acid and H_2O_2	• High compressive strength • Increased porosity	(Kaewmee et al. 2020)
Geopolymerization using alkali activator	Incinerated bottom ash	$NaOH + Na_2SiO_3$	-	• Increased porosity	(Z. Chen et al. 2016)
Geopolymerization using acidic activation	Metakaolin	Phosphoric acid	-	• Acidic geopolymer • High compressive strength	(L. Gao et al. 2020)
Reactive emulsion templating	Metakaolinite	K_2SiO_3	Waste vegetable oil	• Increased pore volume • Increased pore width • Increased surface area	(Medpelli, Seo, and Seo 2014)
Reactive emulsion templating	Metakaolinite	K_2SiO_3	Paraffin	• Increased pore volume • Increased pore width • Reduced • surface area	(Medpelli, Seo, and Seo 2014)
Reactive emulsion templating	Metakaolinite	K_2SiO_3	Canola oil	• Reduced pore volume • Increased pore width • Reduced • surface area	(Medpelli, Seo, and Seo 2014)
Direct foaming	Fly ash	$NaOH + Na_2SiO_3$	Aluminum powder	• Pore size enlargement	(Sanjayan et al. 2015)
Direct foaming	Metakaolin	KOH/K_2SiO_3	Silicon powder	• Pore size enlargement	(Medri et al. 2013)
Direct foaming	Metakaolin	KOH/K_2SiO_3	Hydrogen peroxide+ egg white	• Enhanced porosity	(Bai and Colombo 2017)
Direct foaming	Fly ash	NaOH	NaOCl	• Enhanced porosity	(Böke et al. 2015)
Desilication	Fly ash	$NaOH + Na_2SiO_3$	NaOH	• Enhanced surface area	(Kumari et al. 2021)

Modification	Precursor	Activator	Modifying agent	Modified properties	Reference
Ion exchange+ Dealumination+ Desilication	Halloysite clay	NaOH + Na_2SiO_3/ KOH + K_2SiO_3	Na_2H_2 EDTA NaOH	• Enhanced acidity • Enhanced pore volume • Enhanced surface area	(Alzeer, MacKenzie, and Keyzers 2016)
Oil templating + foaming	Metakaolin	KOH + K_2SiO_3	Hydrogen peroxide	• Enhanced porosity	(Cilla, Morelli, and Colombo 2014)
Ion exchange with ammonium nitrate	Metakaolin	Water glass+NaOH	Ammonium nitrate	• Almost 98% ion-exchange was accomplished	(Sazama et al. 2011)
Ion exchange with ammonium nitrate	Metakaolin	Potassium water glass	Ammonium nitrate	• Almost 98% ion-exchange was accomplished	(Sazama et al. 2011)
Ion exchange with Cu	Metakaolin	Water glass+NaOH	Cupric nitrate	• Almost 98% ion-exchange was accomplished	(Sazama et al. 2011)
Ion exchange with Cu	Metakaolin	Potassium water glass	Cupric nitrate	• Almost 90% ion-exchange was accomplished	(Sazama et al. 2011)
Ion exchange with Co	Metakaolin	Potassium water glass	Cobalt nitrate	• 80% Ca^{2+} and K+ exchange was accomplished	(Sazama et al. 2011)
Ion exchange with NH_4 Cl	Metakaolin	NaOH + Na_2SiO_3	Ammonium chloride	• 100% ion exchange was accomplished • Pore diameter was reduced • Surface area was increased	(José Ramón Gasca-Tirado 2018)
Ion exchange with $(NH_4)_2TiO_2(C_2O_4)_2$	Metakaolin	NaOH + Na_2SiO_3	Ammonium oxodioxalatotitanate (IV)	• 100% ion exchange was accomplished • Pore diameter was reduced Surface area was increased	(José Ramón Gasca-Tirado 2018)
Ion exchange with $(CH_3)_4NBr$	Metakaolin	NaOH + Na_2SiO_3	Tetramethylammonium chloride	• 100% ion exchange was accomplished • Pore diameter was reduced • Surface area was increased	(José Ramón Gasca-Tirado 2018)

Table 2. (Continued)

Modification	Precursor	Activator	Modifying agent	Modified properties	Reference
Fe loaded catalyst	Metakaolin	$NaOH + Na_2SiO_3$	Ferric chloride	• Highly dispersed Fe species in the form of Fe(III) complexes) were formed	(Sazama et al. 2011)
Pt loaded catalyst	Metakaolin	$NaOH + Na_2SiO_3$	trans-Diamminetetrachloroplatinum (IV)	• Clusters of platinum species of undefined sizes	(Sazama et al. 2011)
Replica route	Calcined kaolin clay and precipitated silica and Chamotte	K_2SiO_3	Polyurethane	• Highly porous geopolymer was obtained	(Kovářík et al. 2017)
Ice templating	Metakaolin	K_2SiO_3	Water	• Hierarchical geopolymers with mesoporous matrices and lamellar macro-porosities • High surface area geopolymer	(Papa et al. 2015)
Additive manufacturing	Metakaolin	$NaOH + Na_2SiO_3$	Polyethylene glycol	• Enhanced porosity • Low surface area • Large pore size	(Franchin et al. 2020)
Loading	Fly ash	$KOH + Na_2SiO_3$	C_6H12N_4, NH4Ac, $In(NO_3)_3 \cdot 4.5H_2O$, $Ni(NO_3)_2 \cdot 6H_2O$	• Mesoporous geopolymers with photocatalytic properties were prepared	(Y. J. Zhang, Zhang, Kang, and Zhang 2017)
Composite	Metakaolin	Silica sol + KOH	Graphene oxide	• Flexural strength and fracture toughness were enhanced	(Yan et al. 2015)

Sharma et al. (Sharma et al. 2015) prepared a calcium-loaded geopolymer catalyst for a base-catalyzed transesterification reaction. The geopolymer was made porous by reactive emulsion templating with canola oil, followed by loading calcium ions by incipient wetness impregnation technique. The prepared calcium-loaded geopolymer contained surface Ca sites resulting in high basicity compared to CaO or other alkaline earth metal oxides (Sharma et al. 2015). Preparation and application of bimetallic oxide semiconductors of In and Ni over alkali-activated geopolymer for water splitting reaction was attempted by Zhang et al. (Y. J. Zhang, Zhang, Kang, and Zhang 2017). Hexamethylenetetramine was used as the pore-forming agent during geopolymerization, which was followed by ion exchange and impregnation steps. The catalyst was reported to possess a wide response to the sunlight spectrum but had a weak response to the photoluminescence spectrum. The same team made another attempt towards photocatalytic hydrogen production using an Mn^{2+}-doped and CuO loaded graphene bottom ash-based geopolymeric (Mn^{2+}-CuO/GBAG) composite (Y. J. Zhang, He, et al. 2017). Similar to Zhang et al. (Y. J. Zhang, Zhang, Kang, and Zhang 2017), this catalyst, too, was prepared by alkali activation to prepare geopolymer followed by ion exchange and impregnation steps, but no pore-forming agent was used in this case. Additionally, the prepared catalyst was also deemed suitable for degradation of dye wastewater. Candamano et al. (Candamano et al. 2015) prepared a geopolymer-based catalyst loaded with nickel for steam reforming, partial oxidation, and autothermal steam reforming of ethanol to syngas. This nickel-loaded catalyst was prepared by loading an alkali-activated geopolymer by incipient wetness impregnation with an ethanol solution of nickel salt. The catalyst was reported to yield complete ethanol conversion with the highest amount of hydrogen produced under steam reforming conditions. The geopolymer modification methods, the precursor, activator, modifying agent and the modified properties are summarised in Table 2.

Types and Properties of Geopolymer Based Catalysts

As catalysts, geopolymers can be modified via a range of treatments, including ion exchange, demetallation, and impregnation with metals or specific ions, depending upon the application. All of these treatments impart targeted characteristic changes in the geopolymer, which are discussed here. These modifications can be monitored via FE-SEM, TEM, XRD, XPS, FTIR, XRF, MAS-NMR, UV-DRS, and UV-photoluminescence techniques.

Geopolymers are amorphous aluminosilicates whose amorphicity is confirmed by a broad hump in the range of 20-40° 2θ. This hump represents the amorphous characteristic of practically all poly(sialate-siloxo) and higher sialates resulting from polycondensation of alkali aluminosilicates (Davidovits 2008). This feature, along with the presence of a sodium aluminosilicate at around 26° 2θ, has been reported by several authors (Kumari et al. 2021) (Sazama et al. 2011) (Alzeer and MacKenzie 2018). Apart from this, several sharp signals of quartz, mullite, magnetite, calcite, and merwinite may also appear depending upon the source of geopolymer. Sazama et al. (Sazama et al. 2011) used granulated blast furnace slag for the preparation of geopolymer and reported the presence of sharp signals of calcite and merwinite at various 2θ values. Alzeer et al. (Alzeer, MacKenzie, and Keyzers 2016) prepared a Beckmann rearrangement reaction specific catalyst by a sequential treatment which resulted in the generation of acidic active sites in the geopolymer. The geopolymer catalyst depicted a slight shift in the characteristic broad hump of the geopolymer and was reported to be an indication of washing of some crystalline phases and Al-rich debris during the chosen sequential treatment. A similar shift in the broad hump was noticed by Sharma et al. (Sharma et al. 2015) too, where the authors reported the shift to have resulted from an increment in the interatomic distances due to the incorporation of calcium ions along with additional oxide ions into the geopolymer matrix. XRD analysis can also be used for studying the changes in crystal sizes and the degree of crystallinity. Kumari et al. (Kumari et al. 2021) studied the changes in crystal size, and degree of crystallinity of a geopolymer catalyst upon modification with three staged sequential treatment. They reported a decrease in the crystal size, while the degree of crystallinity was reported to have increased upon modification. Increase in the degree of crystallinity generates several new crystal phases, which take part in the reaction and thus increase the catalytic activity to a large extent (Oliver et al. 2004) (Nicolaides, Sincadu, and Scurrell 2002). Landi et al. (Landi et al. 2013) have reported that the amount of amorphous phase increases and several crystal phases appear on high temperature thermal treatments. The appearance of those new crystal phases can be expected to increase the efficiency of the catalyst.

Porosity and surface area are of vital importance to a catalyst, and the same can be confirmed by SEM, TEM, and BET analyses. The TEM micrographs of the sequentially treated K-geopolymer catalyst prepared by

Alzeer et al. (Alzeer, MacKenzie, and Keyzers 2016) suggest the presence of large mesopores stacked over each other, while the Na-geopolymer catalyst was reported to possess small randomly oriented non-uniform micropores. In another report by the same authors, Na-geopolymer was reported to possess randomly arranged micropores with high surface area and new mesopores (Alzeer, MacKenzie, and Keyzers 2017). The Na-geopolymer catalyst also possessed interparticle voids, resulting in a high surface area of the catalyst confirmed by BET analysis. The presence of interparticle voids was reported in their other paper as well, but the geopolymer prepared in that case originated from a fly ash source and was only ion-exchanged, explaining the comparatively lower surface area (Alzeer and MacKenzie 2018). Calcium-loaded geopolymer prepared by Sharma et al. (Sharma et al. 2015) possessed a hierarchical structure with large macropores scattered over the mesoporous matrix made up of leisurely connected aluminosilicate nanoparticles. The meso- and macropores of the catalyst can be expected to provide better diffusion of the reactant into the catalyst pores, and the presence of a high surface area facilitates better distribution of active sites.

The incorporation of metals or ions into the geopolymer and the form in which they are existing or have bonded with the geopolymer matrix can be studied by XPS, FTIR, MAS NMR spectroscopy, ICP-OES, and UV-Vis-NIR diffuse reflectance spectroscopy. Authors have monitored the formation of a geopolymer with the help of Al- or Si-MAS-NMR spectroscopy. The presence of tetrahedrally co-ordinated Al indicated by chemical shifts at ~60 ppm is an indication of a well-formed geopolymer and has been noticed with the presence of another negligibly shift at ~4 ppm, which indicates the presence of octahedrally co-ordinated Al resulting from the small amount of their unreacted precursors (Sazama et al. 2011) (Alzeer and MacKenzie 2018) (Alzeer, MacKenzie, and Keyzers 2016) (Alzeer, MacKenzie, and Keyzers 2017). Similarly, the presence of tetrahedrally co-ordinated Si can be confirmed by the chemical shift at ~90 ppm, while the peak may shift to ~94 ppm in the case of Si-rich geopolymers (Alzeer, MacKenzie, and Keyzers 2016) (Alzeer, MacKenzie, and Keyzers 2017). Demetallation treatments may result in the shift of the peaks due to changes in the concentration of silanol groups (Alzeer, MacKenzie, and Keyzers 2016). XPS analysis gives an idea of the elemental composition and the chemical and electronic state of the atoms in a material, and the results are represented in the form of intensity versus binding energy spectra. Sharma et al. (Sharma et al. 2015) analyzed their Ca loaded geopolymer with the help of XPS analysis and reported that the calcium loaded onto the geopolymer existed in the form of Ca^{2+} ions, which is not due to the CaO bond but due to the presence of a covalent aluminosilicate network surrounding the Ca ions.

The suitability of a catalyst for any targeted application can be further studied via its acidity or basicity. The basicity of a catalyst can be studied via CO_2-TPD analysis, while NH_3-TPD is used for analyzing the acidity. The TPD profile gives an idea of the types and strength of the acidic or basic sites, and the integration of the total area gives the amount of CO_2 or NH_3 desorbed from the catalyst surface. Sharma et al. (Sharma et al. 2015) reported the basicity of their Ca-loaded geopolymer catalyst to be 0.77 mmol/g, which, as reported, was substantially higher than other CaO or alkaline earth metal oxides, and this basicity was directly related to the biodiesel yield.

The suitability for any targeted application can also be understood with the help of functional groups present on a catalyst surface, which can be identified by FTIR spectroscopy. Geopolymers show some characteristic geopolymer bands at around 1000 and 1084 cm^{-1}, which correspond to Si-O-Al and Si-O-Si streching vibrations, respectively (Alzeer, MacKenzie, and Keyzers 2017) (Alzeer and MacKenzie 2018; Alzeer, MacKenzie, and Keyzers 2016) (Kumari et al. 2021). Geopolymers also show a broad peak at around 3400-3500 cm^{-1}, which are ascribed to H-bonded silanol nests (Alzeer, MacKenzie, and Keyzers 2017) (Alzeer and MacKenzie 2018; Alzeer, MacKenzie, and Keyzers 2016) (Kumari et al. 2021). Another peak corresponding to adsorbed water molecules appears at around 1640 cm^{-1} (Alzeer, MacKenzie, and Keyzers 2017) (Alzeer and MacKenzie 2018; Alzeer, MacKenzie, and Keyzers 2016) (Kumari et al. 2021). Sharma et al. (Sharma et al. 2015) have confirmed the presence of adsorbed water molecules through their TGA thermograms. In some cases, CO_2 asymmetric vibrations due to atmospheric carbonation of geopolymer may also appear at around 1480 and 2347cm^{-1}. The generation of acidity via ion-exchange treatment for conversion of Na-form of geopolymer to its H-form can be monitored by peaks at 3700 and 3620 cm^{-1}, which correspond to H-bonded vicinal silanols and Bronsted bridging hydroxyls, respectively (Alzeer, MacKenzie, and Keyzers 2017) (Alzeer and MacKenzie 2018; Alzeer, MacKenzie, and Keyzers 2016) (Kumari et al. 2021). Alzeer et al. (Alzeer and MacKenzie 2018) have also reported the presence of bands at 1445, 1600, 1620, and 1545 cm^{-1} corresponding to Lewis and Bronsted acid sites. The catalytic properties of geopolymers can be modified and monitored with the above-discussed methods. These methods also aid in understanding the mechanism of the reaction they are subjected to. The utilization of geopolymers in relevant catalytic applications is discussed in the next section.

Applications of Geopolymer Based Catalysts

Geopolymers have found wide usage in adsorption; however, their usage as catalysts is comparatively limited to specific areas. They can be used as catalysts to eliminate pollutants and improve reaction yield. They have been used in the production of biodiesel and hydrogen, for degradation of dyes, purification of gases, and petroleum refining.

Geopolymer Based Catalysts for Biodiesel Production

In industries, transesterification of triglycerides in oils and fats is usually conducted via homogenous catalysis. Although the esterification reaction is completed without difficulty due to the miscible catalyst, removal from the product is a difficult task, and corrosion of the reactor surfaces by the catalysts is another undesirable factor. The need for heterogeneous catalysts is growing because of their economic benefits, such as inexpensive preparation, high thermal stability, low corrosion effect, easy separation from the reaction mixture, and reusability (Berrios and Skelton 2008; Mardhiah et al. 2017).

Transesterification reactions can be carried over both acid or base catalysts, but base-catalyzed transesterification reactions are faster than acid-catalyzed ones (Derrien, Renard, and Brunei 1998). Hence, inherently basic geopolymers (although acidic geopolymers also exist) can be a potential catalyst for transesterification. Derrien et al. (Derrien, Renard, and Brunei 1998) suggest that transesterification reaction works best with strongly basic and porous catalytic systems. Literature suggests that geopolymer has been highly utilized towards transesterification. Esterification of soybean oil with CaO-loaded geopolymer, sodium, and potassium-based geopolymers is common. These reactions require the presence of methanol and can yield conversions as high as 100% (Sharma et al. 2015). The lattice-shaped Na-geopolymer catalyst could yield 43.5% of biodiesel, while the ground powdered form could yield 51.4% biodiesel formation with 150 mol% excess methanol (Innocentini et al. 2019). In another study, sodium and potassium-based geopolymer powders were used as the catalyst, but they also used 150 mol% excess methanol, which is a major setback of this process (Botti et al. 2020). Na-based catalyst was reported to perform better with fatty acid methyl ester content in the biodiesel phase at 85.1% and 89.9%, respectively. The effects of sodium and potassium on the geopolymerization process and the accessibility of the reactants to the catalytic sites was responsible for the differences in the performance (Botti et al. 2020).

Apart from soybean oil, waste cooking oil could also serve as a potential source for biodiesel formation. A potassium geopolymer catalyst was used for the transesterification reaction of sunflower waste cooking oil (SWC). The catalyst was prepared using natural metakaolin and synthetic Mg/Al LDH and MCM-41 loaded with potassium. Similar to previous reports, they also used excess methanol (methanol to oil ratio of 16.4:1) and achieved an yield of 94.6% (Sayed et al. 2020). In another study, a rice husk ash-based geopolymer was used to synthesize biodiesel from *Calophyllum inophyllum* seed oil, but the yield was the lowest (87%) of all the reports in this section (Saputra et al. 2018).

Geopolymer Based Catalysts for Hydrogen Production

Combustion of fossil fuels has caused numerous environmental problems, including air pollution. An increase in energy demand has created a strong need for hydrogen production as hydrogen is the source of a clean and highly efficient form of energy. One of the best ways to produce hydrogen is water splitting by photocatalytic mechanism (Matsuoka et al. 2007). Traditionally, transition metal catalysts like Mn, Ni, Co, and Fe and photocatalysts such as CdS, TiO_2, and SiC have found wide application in water splitting reactions.

Geopolymers can be synthesized via a wide range of waste materials, and those wastes often contain valuable entities. Granular blast furnace slag contains natural oxide semiconductors like Fe_2O_3 and TiO_2, which act as active species after a successful geopolymerization. One such alkali-activated slag-based geopolymer was prepared by Zhang et al. (Y. J. Zhang and Chai 2014). The prepared geopolymer possessed a mesoporous structure and contained Fe_2O_3 and TiO_2 semiconductors in their structure matrix. Those acted as the active species for the reaction and the combined action of the mesoporous structure, and the presence of active species yielded 51697 μmol/g of H_2 after 6 h of reaction under a 500W xenon lamp as a light source. The same team synthesized a $CaWO_4$-loaded granular blast furnace slag-based geopolymer by ion exchange with NH_4Ac followed by loading with $(NH_4)_6W_7O_{24}\cdot 6H_2O$ (Y. J. Zhang, Zhang, Kang, Yang, et al. 2017). The active phase of $CaWO_4$ was formed by the interaction between $(NH_4)_6W_7O_{24}\cdot 6H_2O$ and calcium ions present in the slag. Similar to the results obtained in their previous report, the highest yield of H_2 was 50827 μL/g under xenon lamp irradiation for 6 h. The catalyst also contained active components of TiO_2 and the matrix of calcium silicate hydrate, and the high yield was due to the synergistic effect of the active components of $CaWO_4$, TiO_2, and the matrix of calcium silicate hydrate.

Modified or metal-loaded geopolymers also serve as good catalysts for water splitting. A nanocatalyst prepared by loading In_2O_3 and NiO into geopolymer prepared from fly ash precursor was used to produce hydrogen by water splitting by Zhang et al. (Y. J. Zhang, Zhang, Kang, and Zhang 2017). As reported, the photocatalytic activity of H_2 production depended on the formation and separation of the photo-generated carriers on active components of the In_2O_3 and NiO. They prepared a range of catalysts and checked their activity towards water splitting, and found that the catalyst with the highest BET specific surface area of 21.87 m^2/g, the largest mesoporous volume of 0.0758 mL/g and the weakest photoluminescence intensity as well as appropriate band gap of 2.84 eV, demonstrated the highest photocatalytic activity (Y. J. Zhang, Zhang, Kang, and Zhang 2017).

Graphene bottom ash is a good source of quartz and anorthite. A geopolymer composite synthesized by Zhang et al. (Y. J. Zhang, He, et al. 2017) by alkaline activation of graphene bottom ash primarily contained amorphous geopolymer, graphene, and minerals of quartz and anorthite. The authors modified the composite by doping with Mn^{2+} and loading with CuO and used it in photocatalytic water splitting under a xenon lamp. The hydrogen generation was reported to increase with the increase in doped Mn^{2+} and loaded CuO contents. The Mn^{2+}-20CuO-GBAG composite displayed the highest H_2 yield, which was due to the weakest photoluminescence, which correlates to the high reduction efficiency of photoelectrons (Y. J. Zhang, He, et al. 2017). Graphene can also be used to prepare a graphene reinforced blast furnace slag-based nanocomposite for photocatalytic hydrogen production under a xenon lamp. The hydrogen production yield was reported to increase with an increasing amount of graphene due to the synergic effect of graphene and TiO_2 and Fe_2O_3 from the alkali-activated blast furnace slag (Y. J. Zhang, Zhang, and Yang 2018). In another study of the same kind, the authors prepared a CeO_2-loaded porous alkali-activated steel slag-based photocatalyst and achieved optimum hydrogen production yield after 5 h of reaction time (Kang et al. 2017). The hydrogen generation was reported to occur due to the presence of active species of the CeO_2 and FeO in the carrier (Kang et al. 2017).

Other than photocatalytic degradation of water, hydrogen can also be produced using geopolymer catalyst by reforming processes such as steam reforming (SR), autothermal steam reforming (ATR), and partial oxidation (POX). In one study, Ni/geopolymer catalyst produced syngas at 700°C in SR, ATR, and POX of ethanol. The catalyst's performance in all three reactions was reported to be stable and almost complete in terms of ethanol conversion. They achieved the largest amount of hydrogen under SR and ATR conditions (70 mol%) and the largest amount of carbon monoxide during POX reactions (Candamano et al. 2015).

Geopolymer Based Catalysts for Photocatalytic Dye Degradation

Photocatalytic dye degradation is another area emerging as a topic of significant interest to the research community. Geopolymer has been extensively utilized as a catalyst in dye degradation. A fly ash-based geopolymer was used as a photocatalyst for methylene blue (MB) degradation from wastewater (Y. Zhang and Liu 2013). The authors employed three methods, viz., direct photolysis, adsorption, and photocatalysis, to achieve optimum dye degradation, and photocatalysis yielded the best removal efficiency. The highest photocatalytic degradation was reported to be a synergistic effect of adsorption and semiconductor photocatalysis (Y. Zhang and Liu 2013). In another study, a granulated blast furnace slag-based geopolymer loaded with Fe_2O_3 was used for photocatalytic degradation of Congo red (CR) dye (Y. J. Zhang et al. 2013). The 5% loaded Fe_2O_3 catalyst was reported to have the highest photocatalytic degradation rate and lowest adsorption intensity in the near infrared region when exposed to UV irradiation. The degradation was reported to be due to the uniform distribution of active Fe_2O_3 clusters on the geopolymer surface, giving easy accessibility to the active sites (Y. J. Zhang et al. 2013). Li et al. (Li et al. 2016) used geopolymer spheres loaded with CdS crystals for photocatalytic degradation of methyl orange with an efficiency of 92.57% due to the combined effects of adsorption and photodegradation (38.85% adsorption and 53.72% photodegradation).

Graphene-based geopolymers have also found wide usage in photocatalytic dye degradation. A geopolymer from graphene bottom ash (GBAG) doped with Mn^{2+} and loaded with CuO was used for the photocatalytic degradation of direct sky blue 5B dye under UV irradiation. The authors achieved complete degradation of the dye with a 15% CuO-loaded geopolymer. This extraordinary efficiency was obtained due to doped Mn^{2+}, which enhances CuO to a donor level, thus increasing the oxidation performance of photo holes (Y. J. Zhang, He, et al. 2017). In another study, graphene was incorporated into a fly ash-based geopolymer to increase the electron conductivity of the geopolymer and used in the degradation of indigo carmine. As reported, the highest degradation efficiency of 90.2% was obtained, with the catalyst having the highest BET surface area, mesoporous volume and pore volume, and smallest average pore size. Graphene reportedly helped in increasing the acceptance rate of electrons generated by Fe_2O_3 in the FAG thereby improving the degradation (Y. J. Zhang et al. 2018). In another study, a graphene alkali-activated steel slag geopolymer loaded with CdO was used for the degradation of DFB dye under UV radiation. The authors achieved complete degradation of dye with 8% CdO loaded geopolymer, which was due to the lower energy gap and incorporation of graphene, which made the excitation of photo-generated electrons and holes simpler (Y. J. Zhang, He, and Chen 2018). In another study where a graphene-geopolymer nanocomposite was used for the degradation of methyl violet dye under UV irradiation, graphene was found to enhance the microstructure, average pore size, pore volume percentage, and BET surface area, thus increasing the degradation rate (Y. J. Zhang et al. 2016).

The photocatalytic activity of TiO_2 is quite popular. A TiO_2 nanoparticles incorporated fly ash geopolymer prepared by Maity et al. (Maiti et al. 2020) was used to study the photocatalytic oxidation of methylene blue under UV irradiation. The photocatalytic degradation rate was reported to increase with loading in each catalyst and was inversely proportional to the particle size of TiO_2 NPs (Maiti et al. 2020). In another study, a photocatalyst prepared by coating nanotitania in metakaolin and rice husk ash-based geopolymer spheres was used for the degradation of dyes. The catalyst demonstrated 90% methylene blue degradation under UV irradiation (Isabel Bravo et al. 2019).

In some other studies, electroconductive geopolymer composites were prepared for photocatalytic de degradation. An electroconductive geopolymer composite fabricated by coupling $CaMoO_4$, electroconductive carbon black, and silicomanganese slag-based geopolymer prepared by He et al. (He et al. 2019) was used to study the degradation activity towards the degradation of basic violet 5BN dye. The authors achieved complete degradation with 0.54 g of 3.78 wt% $CaMoO_4$ loaded catalyst (He et al. 2019). Another electroconductive geopolymer composite was synthesized by the same authors using geopolymer based on circulating fluidized bed fly ash and carbon fiber and used in the degradation of basic violet 5BN (He et al. 2020). The introduction of carbon fiber was reported to increase the electroconductivity, which further enhanced the separation of photo-generated electron-hole, which ultimately increased the photocatalytic activity (He et al. 2020). Apart from these, a perlite-based geopolymer synthesized using perlite was reported to effectively degrade methylene blue from an aqueous solution (Saufi et al. 2020). The authors reported that when the geopolymer was used under UV irradiation, the degradation was more than that of using geopolymer without UV radiation or using UV radiation without geopolymer. This was reported to appear as a result of the synergic effect of photocatalytic degradation and adsorption (Saufi et al. 2020).

Geopolymer Based Catalysts for Other Catalytic Applications

Heterogeneous catalysts are more economical and environmentally friendly as compared to homogenous catalysts. Nowadays, geopolymer-based catalysts are being used for varied catalytic applications. A geopolymer-based catalyst prepared by Alzeer et al. (Alzeer, MacKenzie, and Keyzers 2016) was used for the Beckmann rearrangement of cyclohexane oxime to ε-caprolactum. The authors reported that the catalyst with a small surface area and large pore diameter showed excellent conversion of oxime but less selectivity towards the caprolactum product. They reported that the catalyst's surface area and pore diameter could be changed by varying the initial composition of inorganic polymers. They increased the surface area and enhanced the distribution of the active sites by exposing the catalyst to post-synthetic dealumination and desilication. The catalytic activity and performance of the geopolymeric catalyst was reported to be superior to other solid catalysts under the same reaction conditions and could be reused for up to 5 cycles without significant activity loss (Alzeer, MacKenzie, and Keyzers 2016). Using halloysite clay as a geopolymer precursor, the same authors prepared another geopolymer catalyst via ion exchange with NH_4^+ (Alzeer, MacKenzie, and Keyzers 2017). They subjected the catalyst to Friedel-Crafts alkylation reactions of benzene, anisole, p-xylene, mesitylene, and toluene. The starting compositions directly affected the acidity and porosity of the catalyst, which were enhanced after sequential desilication and dealumination treatments (Alzeer, MacKenzie, and Keyzers 2017). In another study, the same authors prepared a similar catalyst with fly ash as the precursor, and the post-synthetic modification was done by ion-exchange with NH_4^+ (Alzeer and MacKenzie 2018). The catalyst was used in the Friedel-Crafts benzylation reaction of benzene, anisole, p-xylene, mesitylene, and toluene with benzyl halide. The catalysts in which fly ash containing higher Fe_2O_3 contents were used as precursors were reported to be more reactive than those with lower Fe_2O_3 contents. Also, those catalysts reportedly had higher reactivity than Fe-ZSM-5, fly ash supported sulphated zirconia and ionic liquids (Alzeer and MacKenzie 2018). They also synthesised geopolymers with C-class fly ash or high Ca containing coal fly ash precursors and used them in Friedel-Crafts acylation reaction of anisole, p-xylene, mesitylene and toluene using benzyl chloride (BzCl). They achieved a conversion of almost 100% in case of anisole, ~95% for mesitylene, ~14% for toluene and 30% for p-xylene. The reactivity of the catalyst is due to Lewis and Bronsted acidic sites and due to the Fe_2O_3 species in the precursors (Alzeer and Mackenzie 2019).

The application of geopolymeric catalyst could also be extended to isopropanol dehydrogenation. A geopolymer-supported nano bimetallic catalyst loaded with Cu and Ni was synthesized using metakaolin as a geopolymer precursor for isopropanol dehydrogenation (John, Hu, and Zhu 2018). The catalyst yielded a maximum conversion of 23.18% and acetone selectivity of 88.24%. They also reported that the catalyst could be reused for up to 5 cycles; however, the low conversion efficacy indicated that the catalyst was not suitable for this application (John, Hu, and Zhu 2018). In another study, a geopolymer synthesized using metakaolin and waste silica fumes as precursors and modified with active carbon was used for NH_3-SCR reaction (H. Chen et al. 2020). The catalyst was prepared by the introduction of iron cations by incipient wet impregnation and calcination, wherein an excess of cations were captured by the activated carbon and was partially reduced. Fe ions were reported to be highly dispersed in geopolymer and activated carbon, and when the loading was less, monomeric isolated $[Fe(OH)]^{2+}$ and oligomeric FeO_X clusters were found to form in the geopolymer ring and activated carbon, respectively. As the iron loading increased, the isolated monomeric $[Fe(OH)]^{2+}$ was reportedly closer and formed less bonded dimeric $[HO\text{-}Fe\text{-}O\text{-}Fe\text{-}OH]^{2+}$ due to internal condensation. This increase in iron loading in the composite was reported to increase the selectivity and reactivity (H. Chen et al. 2020). The study of the activity of geopolymeric catalyst towards NH_3-SCR was studied by Sazama et al. (Sazama et al. 2011) too. The authors prepared a metakaolin-based geopolymer, which was modified via ion exchange with NH_4^+ and then with Co, Cu, and Pt species and separately loaded with Fe species via impregnation. They used the catalysts for the catalytic oxidation of volatile organic compounds and in the NH_3-SCR-NO_X reaction. Pt, Co, and Fe incorporated geopolymer catalysts were used to oxidize decane. The conversion to CO_x and selectivity towards CO_2 and CO was reported to be dependent on temperature, while the Pt geopolymer, even at low temperature, yielded high conversion, and almost half conversion of decane was reached at 135°C. They noticed that in the case of Co and Fe geopolymer catalysts, the reaction began at 260°C and total conversion was obtained at around 470-480°C. The selectivity of CO_2 in these cases was reported to increase with increasing temperature. The Cu and NH_4^+ geopolymers were used for the NO_X and NH_3 conversion in the NH_3-SCR-NO_X reaction and could yield better conversion than commercially available V_2O_5/Al_2O_3 even in the temperature range of 250-350°C (Sazama et al. 2011).

A TiO_2 incorporated metakaolin geopolymer prepared via ion exchange was used for the photolysis of 2-butanone into acetone and acetaldehyde (J. R. Gasca-Tirado et al. 2014). The presence of TiO_2 bought significant physical changes in the geopolymer, and 2-butanone was degraded at TiO_2 active sites for the adsorption/desorption of reactants and products, respectively.

In another study, TiO_2-based geopolymers were used for the photocatalytic degradation of nitric oxide (Strini et al. 2016). The authors prepared geopolymers using four different precursors, viz., metakaolin, fly ash, hybrid siloxane-metakaolin, and foamed hybrid siloxane-metakaolin, and loaded each with 3% by weight of TiO_2. The photocatalytic activity was reported to be dependent on the precursors, with the degradation rate of NO ranging from 3-0 nmol/m^2.s, and for NO_X, the rate was reported to be comparatively low (Strini et al. 2016). As Bendoni et al. (Bendoni et al. 2019) reported, geopolymers could also be used for catalytic tar decomposition. To achieve that, they prepared a Fe_2O_3 and Mn_2O_3 loaded geopolymer matrix and compared the activity of these catalysts with Olivine (natural magnesium-iron silicate), a reference catalyst. As reported, Olivine increased tar decomposition from 33% at 700°C to 64% at 900°C, whereas geopolymer catalysts showed an increase in tar decomposition over 70% at 700°C, which slightly increased after 800°C to decrease again at 900°C. Compared to Olivine, which is nonporous with no internal surface, geopolymer catalysts had a well-defined porous structure and high surface area. The oxide phases are only embedded in the geopolymer matrix and are higher in amount, whereas in Olivine, Fe_2O_3 is in the crystal structure of the mineral. These factors reflected why the geopolymer catalyst performed better than Olivine. Among the geopolymer catalyst, the one which contained both Mn^{2+} and Fe^{2+} performed better due to the synergetic effect of both the metal oxides (Bendoni et al. 2019).

In addition to these, geopolymers could also be modified into efficient catalysts for some new applications like CO_2 methanation, petroleum refining, tetracycline degradation, dichloromethane oxidation, and naphthalene degradation. Seelam et al. (Seelam et al. 2021) prepared a high surface area geopolymer with a metakaolin source and used it to support the oxidation of dichloromethane. The modified geopolymer, prepared by treatment with hydrochloric acid, possessed a high specific surface area and yielded a 90% conversion of dichloromethane. The authors reported that total oxidation of dichloromethane is usually possible with noble metals, and the high efficacy obtained by geopolymer prepared without any loading indicated that they are promising support for catalytic applications. Kumari et al. (Kumari et al. 2021) prepared a sulfated geopolymer catalyst by sequentially treating the geopolymer with desilication, ion exchange, and sulfation steps and subjected the prepared catalyst to the oxidation of dibenzothiophene. The authors obtained 85% conversion of geopolymer without using any metal loading. Li et al. (J. Huang et al. 2020) prepared a geopolymer-biochar composite by in-situ synchronous carbonation and self-activation process. The composite reportedly exhibited a porous structure and yielded 100% tetracycline conversion. The authors reported that the catalyst demonstrated good stability, reusability, and potential to activate hydrogen peroxide, which ultimately provided a good activity for a Fenton-like reaction. In another study, a similar biochar-geopolymer composite catalyst loaded with Cu and Fe nanoparticles was prepared and subjected to the degradation of naphthalene from water and soil via activation of potassium peroxymonosulfate (Zhu et al. 2022). As reported, the catalyst could degrade 98.35% and 67.98% of naphthalene present in water and soil and could be effectively reused for up to 4 cycles. In a recent study, Wan et al. (Wan et al. 2022) prepared nickel-loaded slag-based geopolymer spheres for CO_2 methanation and reported that the catalyst exhibited 99.2% selectivity for CH_4 and yielded 80.2% conversion of CO_2. The conversion efficiency obtained in this work was reportedly higher than other nickel-loaded catalysts prepared by other authors. These studies indicate the growing interest towards geopolymeric catalysts and demonstrate their feasibility in various catalytic applications. The application of other kinds of geopolymeric catalysts is expected to increase soon.

Conclusion

Geopolymers are emerging technology in the field of sustainable development and environmental remediation. They can be synthesized by a wide range of wastes, including industrial, natural, and agricultural wastes too. This makes them potential alternatives to cease or minimize the pollution caused by the piling or hazardous dumping of these wastes. The utilization of waste materials reduces the overall production cost, and the valuable entities present in them often become sources of desired entities suitable for specific applications. Their properties can be tailored by changing small parameters, and their thermal stability makes them suitable for usage in high-temperature catalytic applications. This review focused on the catalytic applications of geopolymers, prepared mainly from a range of waste materials. The usage of geopolymers in biodiesel production, hydrogen production, and photocatalytic dye degradation has recently gained huge popularity. Their catalytic applications have extended to other oxidative reactions such as dibenzothiophene oxidation, dichloromethane oxidation, and tetracycline oxidation. One of the most recent extensions of geopolymeric catalysts is in the field of CO_2 methanation. In all the cases, geopolymers have exhibited excellent efficiency and thus can be regarded as potential catalysts for more catalytic applications.

Future Perspectives

Geopolymers have served as excellent adsorbents for varied adsorptive applications and have also gained huge popularity in the field of catalysis. Their similarity with zeolites and ability to undergo ion exchange makes them capable of trapping various elements into their matrix, thereby making it possible to modify them according to the desired application conveniently. However, the ability of geopolymer to trap any ion depends upon the size of the cavity, which is not yet clearly studied. An insight into the size of the cavity and its ability to trap specific ions can make it convenient for future researchers to carry their research in that direction. While some discussion on this is available in a few works of literature, conducting experimental studies with all possible combinations to gain more clarity can make the studies costlier and are limited only to the knowledge of the ion-exchange capability of geopolymers. Hence, computational studies aimed at the study of the geopolymer matrix and the size of its cavity can be of substantial help.

Further, geopolymers can be synthesized from a wide range of sources, and kitchen and electronic wastes are other viable resources containing abundant recyclable entities. The key issue in preparing geopolymeric catalysts or any other heterogenous catalyst is fabricating thermally stable, highly active, inexpensive, and recyclable catalysts. Therefore, further studies towards preparing sustainable and eco-friendly geopolymeric catalysts and their applications towards more catalytic applications are required.

References

Alzeer, Mohammad I. M., and Kenneth J. D. Mackenzie. (2019). Fly Ash-Based Geopolymers as Sustainable Bifunctional Heterogeneous Catalysts and Their Reactivity in Friedel-Crafts Acylation Reactions. *Catalysts* 9 (4). https://doi.org/10.3390/catal9040372.

Alzeer, Mohammad I. M., and Kenneth J. D. MacKenzie. (2018). Synthesis and Catalytic Properties of New Sustainable Aluminosilicate Heterogeneous Catalysts Derived from Fly Ash. *ACS Sustainable Chemistry and Engineering* 6 (4): 5273–82. https://doi.org/10.1021/acssuschemeng.7b04923.

Alzeer, Mohammad I. M., Kenneth J. D. MacKenzie, and Robert A. Keyzers. (2016). Porous Aluminosilicate Inorganic Polymers (Geopolymers): A New Class of Environmentally Benign Heterogeneous Solid Acid Catalysts. *Applied Catalysis A: General* 524: 173–81. https://doi.org/10.1016/j.apcata.2016.06.024.

Alzeer, Mohammad I. M., Kenneth J. D. MacKenzie, and Robert A. Keyzers. (2017). Facile Synthesis of New Hierarchical Aluminosilicate Inorganic Polymer Solid Acids and Their Catalytic Performance in Alkylation Reactions. *Microporous and Mesoporous Materials* 241: 316–25. https://doi.org/10.1016/j.micromeso.2016.12.018.

Bai, Chengying, and Paolo Colombo. (2017). High-Porosity Geopolymer Membrane Supports by Peroxide Route with the Addition of Egg White as Surfactant. *Ceramics International* 43 (2): 2267–73. https://doi.org/10.1016/j.ceramint.2016.10.205.

Bai, Chengying, and Paolo Colombo. (2018). Processing, Properties and Applications of Highly Porous Geopolymers: A Review. *Ceramics International* 44 (14): 16103–18. https://doi.org/10.1016/j.ceramint.2018.05.219.

Bai, Chengying, Tao Ni, Qiaoling Wang, Hongqiang Li, and Paolo Colombo. (2018). Porosity, Mechanical and Insulating Properties of Geopolymer Foams Using Vegetable Oil as the Stabilizing Agent. *Journal of the European Ceramic Society* 38 (2): 799–805. https://doi.org/10.1016/j.jeurceramsoc.2017.09.021.

Bajare, Diana, Girts Bumanis, and Aleksandrs Korjakins. (2014). New Porous Material Made from Industrial and Municipal Waste for Building Application. *Materials Science* 20 (3): 333–38.

Barrie, E., V. Cappuyns, E. Vassilieva, R. Adriaens, S. Hollanders, D. Garcés, C. Paredes, Y. Pontikes, J. Elsen, and L. Machiels. (2015). Potential of Inorganic Polymers (Geopolymers) Made of Halloysite and Volcanic Glass for the Immobilisation of Tailings from Gold Extraction in Ecuador. *Applied Clay Science* 109–110: 95–106. https://doi.org/10.1016/j.clay.2015.02.025.

Bendoni, R., F. Miccio, V. Medri, P. Benito, A. Vaccari, and E. Landi. (2019). Geopolymer Composites for the Catalytic Cleaning of Tar in Biomass-Derived Gas. *Renewable Energy* 131: 1107–16. https://doi.org/10.1016/j.renene.2018.08.067.

Berrios, M., and R. L. Skelton. (2008). Comparison of Purification Methods for Biodiesel. *Chemical Engineering Journal* 144 (3): 459–65. https://doi.org/10.1016/j.cej.2008.07.019.

Böke, Nuran, Grant D. Birch, Sammy M. Nyale, and Leslie F. Petrik. (2015). New Synthesis Method for the Production of Coal Fly Ash-Based Foamed Geopolymers.

Construction and Building Materials 75: 189–99. https://doi.org/10.1016/j.conbuildmat.2014.07.041.

Botti, Renata F., Murilo D. M. Innocentini, Thais A. Faleiros, Murilo F. Mello, Danilo L. Flumignan, Leticia K. Santos, Giorgia Franchin, and Paolo Colombo. (2020). Biodiesel Processing Using Sodium and Potassium Geopolymer Powders as Heterogeneous Catalysts. *Molecules* 25 (12): 1–14. https://doi.org/10.3390/molecules25122839.

Buchwald, A., M. Hohmann, K. Posern, and E. Brendler. (2009). The Suitability of Thermally Activated Illite/Smectite Clay as Raw Material for Geopolymer Binders. *Applied Clay Science* 46 (3): 300–304. https://doi.org/10.1016/j.clay.2009.08.026.

Candamano, S., P. Frontera, A. Macario, F. Crea, J. B. Nagy, and P. L. Antonucci. (2015). Preparation and Characterization of Active Ni-Supported Catalyst for Syngas Production. *Chemical Engineering Research and Design* 96: 78–86. https://doi.org/10.1016/j.cherd.2015.02.011.

Chen, Hao, Yao Jun Zhang, Pan Yang He, Chan Juan Li, and Li Cai Liu. (2020). Facile Synthesis of Cost-Effective Iron Enhanced Hetero-Structure Activated Carbon/Geopolymer Composite Catalyst for NH_3-SCR: Insight into the Role of Iron Species. *Applied Catalysis A: General* 605 (13): 117804. https://doi.org/10.1016/j.apcata.2020.117804.

Chen, Zhitao, Yiquan Liu, Weiping Zhu, and En-Hua Yang. (2016). Incinerator Bottom Ash (IBA) Aerated Geopolymer. *Construction and Building Materials* 112: 1025–31. https://doi.org/10.1016/j.conbuildmat.2016.02.164.

Cheng, T. W., and J. P. Chiu. (2003). Fire-Resistant Geopolymer Produced by Granulated Blast Furnace Slag. *Minerals Engineering* 16 (3): 205–10. https://doi.org/10.1016/S0892-6875(03)00008-6.

Cilla, Marcelo Strozi, Márcio Raymundo Morelli, and Paolo Colombo. (2014). Open Cell Geopolymer Foams by a Novel Saponification/Peroxide/Gelcasting Combined Route. *Journal of the European Ceramic Society* 34 (12): 3133–37. https://doi.org/10.1016/j.jeurceramsoc.2014.04.001.

Cong, Peiliang, and Yaqian Cheng. (2021). Advances in Geopolymer Materials: A Comprehensive Review. *Journal of Traffic and Transportation Engineering (English Edition)* 8 (3): 283–314. https://doi.org/10.1016/j.jtte.2021.03.004.

Davidovits, Joseph. (2008). *Geopolymer Chemistry and Applications*. Vol. 171.

Derrien, A., G. Renard, and D. Brunei. (1998). Guanidine Linked to Miceile-Templated Mesoporous Silicates as Base Catalyst for Transesterification. In *Mesoporous Molecular Sieves 1998*, edited by L. Bonneviot, F. Béland, C. Danumah, S. Giasson, and S B T - Studies in Surface Science and Catalysis Kaliaguine, 117:445–52. Elsevier. https://doi.org/10.1016/S0167-2991(98)81023-9.

Duan, Ping, Chunjie Yan, and Wenjun Luo. (2016). A Novel Waterproof, Fast Setting and High Early Strength Repair Material Derived from Metakaolin Geopolymer. *Construction and Building Materials* 124: 69–73. https://doi.org/10.1016/j.conbuildmat.2016.07.058.

Duxson, P., A. Fernández-Jiménez, J. L. Provis, G. C. Lukey, A. Palomo, and J. S. J. Van Deventer. (2007). Geopolymer Technology: The Current State of the Art. *Journal of Materials Science* 42 (9): 2917–33. https://doi.org/10.1007/s10853-006-0637-z.

Esmaily, H., and H. Nuranian. (2012). Non-Autoclaved High Strength Cellular Concrete from Alkali Activated Slag. *Construction and Building Materials* 26 (1): 200–206. https://doi.org/10.1016/j.conbuildmat.2011.06.010.

Franchin, Giorgia, Janne Pesonen, Tero Luukkonen, Chengying Bai, Paolo Scanferla, Renata Botti, Sara Carturan, Murilo Innocentini, and Paolo Colombo. (2020). Removal of Ammonium from Wastewater with Geopolymer Sorbents Fabricated via Additive Manufacturing. *Materials & Design* 195: 109006. https://doi.org/10.1016/j.matdes.2020.109006.

František, Škvára, Šulc Rostislav, T. Zdeněk, S. Petr, Š. Vít, and Z. C. Zuzana. (2014). Preparation and Properties of Fly Ashbased Geopolymer Foams. *Ceramics-Silikáty* 58 (3): 188–97.

Gao, Huan, Hao Liu, Libing Liao, Lefu Mei, Fan Zhang, Longhui Zhang, Sha Li, and Guocheng Lv. (2020). A Bifunctional Hierarchical Porous Kaolinite Geopolymer with Good Performance in Thermal and Sound Insulation. *Construction and Building Materials* 251: 118888. https://doi.org/10.1016/j.conbuildmat.2020.118888.

Gao, Li, Youxiong Zheng, Yan Tang, Jianwei Yu, Xingchang Yu, and Bingxin Liu. (2020). Effect of Phosphoric Acid Content on the Microstructure and Compressive Strength of Phosphoric Acid-Based Metakaolin Geopolymers. *Heliyon* 6 (4): e03853. https://doi.org/10.1016/j.heliyon.2020.e03853.

Gasca-Tirado, J. R., A. Manzano-Ramírez, P. A. Vazquez-Landaverde, E. I. Herrera-Díaz, M. E. Rodríguez-Ugarte, J. C. Rubio-Ávalos, V. Amigó-Borrás, and M. Chávez-Páez. (2014). Ion-Exchanged Geopolymer for Photocatalytic Degradation of a Volatile Organic Compound. *Materials Letters* 134: 222–24. https://doi.org/10.1016/j.matlet.2014.07.090.

Gasca-Tirado, José Ramón. (2018). Ion Exchange in Geopolymers. In, edited by Alejandro Manzano-Ramírez, Ch. 5. Rijeka: IntechOpen. https://doi.org/10.5772/intechopen 80970.

Guisnet, M., and F. Ramôa Ribeiro. (2011). Deactivation and regeneration of solid catalysts. In *Deactivation and Regeneration of Zeolite Catalysts*, Volume 9:3–18. Catalytic Science Series. Imperial College Press. https://doi.org/doi:10.1142/9781848166387_0001.

He, Pan Yang, Yao Jun Zhang, Hao Chen, and Zhi Chao Han. (2020). Development of an Electroconductive Carbon Fiber/Circulating Fluidized Bed Fly Ash Based-Geopolymer Composite for High-Efficiency Treatment of Dye Wastewater. *Ferroelectrics* 565 (1): 1–11. https://doi.org/10.1080/00150193.2020.1761713.

He, Pan Yang, Yao Jun Zhang, Hao Chen, and Li Cai Liu. (2019). Development of an Eco-Efficient $CaMoO_4$/Electroconductive Geopolymer Composite for Recycling Silicomanganese Slag and Degradation of Dye Wastewater. *Journal of Cleaner Production* 208: 1476–87. https://doi.org/10.1016/j.jclepro.2018.10.176.

Hoornweg, Daniel; Bhada-Tata, Perinaz. (2012). What a Waste : A Global Review of Solid Waste Management. World Bank, Washington, DC. https://openknowledge.worldbank.org/handle/10986/17388.

Huang, Jiaqi, Zhili Li, Jiubing Zhang, Yifan Zhang, Yuanyuan Ge, and Xuemin Cui. (2020). *In-Situ* Synchronous Carbonation and Self-Activation of Biochar/Geopolymer Composite Membrane: Enhanced Catalyst for Oxidative Degradation of

Tetracycline in Water. *Chemical Engineering Journal* 397: 125528. https://doi.org/10.1016/j.cej.2020.125528.

Huang, X., L. Yu, Dongwei Li, Yan-Chyuan Shiau, S. Li, and K. Liu. (2015). Preparation and Properties of Geopolymer from Blast Furnace Slag. *Materials Research Innovations* 19 (May). https://doi.org/10.1179/1432891715Z.0000000002210.

Innocentini, Murilo D. M., Renata F. Botti, Paula M. Bassi, Cristina F. P. R. Paschoalato, Danilo L. Flumignan, Giorgia Franchin, and Paolo Colombo. (2019). Lattice-Shaped Geopolymer Catalyst for Biodiesel Synthesis Fabricated by Additive Manufacturing. *Ceramics International* 45 (1): 1443–46. https://doi.org/10.1016/j.ceramint.2018.09.239.

Isabel Bravo, Patricia, Eiza Shimizu, Roy Alvin Malenab, April Anne Tigue, Kimmie Mae Dela Cerna, Jose Isagani Janairo, Michael Angelo Promentilla, and Derrick Ethelbhert Yu. (2019). Nanocrystalline Titania Coated Metakaolin and Rice Hull Ash Based Geopolymer Spheres for Photocatalytic Degradation of Dyes in Wastewater. *Oriental Journal of Chemistry* 35 (1): 167–72. https://doi.org/10.13005/ojc/350118.

Jaimes, J., Angela Montaño, and Claudia González. (2020). Geopolymer Derived from Bentonite: Structural Characterization and Evaluation as a Potential Sorbent of Ammonium in Waters. *Journal of Physics: Conference Series* 1587 (July): 12008. https://doi.org/10.1088/1742-6596/1587/1/012008.

Jiang, Wenguang, Xiangguo Li, Yang Lv, Dongbing Jiang, Zhuolin Liu, and Chenhao He. (2020). Mechanical and Hydration Properties of Low Clinker Cement Containing High Volume Superfine Blast Furnace Slag and Nano Silica. *Construction and Building Materials* 238: 117683. https://doi.org/10.1016/j.conbuildmat.2019.117683.

John, Satmon, Zhengfeng Hu, and Mingqiao Zhu. (2018). Synthesis of Novel Geopolymer Supported Nano Bimetallic Catalysts and Its Application for Isopropanol Dehydrogenation. *Key Engineering Materials* 777: 190–95. https://doi.org/10.4028/www.scientific.net/KEM.777.190.

Kaewmee, Patcharanat, Mengzhu Song, Mitsuyasu Iwanami, Hiroshi Tsutsumi, and Fumitake Takahashi. (2020). Porous and Reusable Potassium-Activated Geopolymer Adsorbent with High Compressive Strength Fabricated from Coal Fly Ash Wastes. *Journal of Cleaner Production* 272: 122617. https://doi.org/10.1016/j.jclepro.2020.122617.

Kang, Le, Yao Jun Zhang, Li Zhang, and Ke Zhang. (2017). Preparation, Characterization and Photocatalytic Activity of Novel CeO_2 Loaded Porous Alkali-Activated Steel Slag-Based Binding Material. *International Journal of Hydrogen Energy* 42 (27): 17341–49. https://doi.org/10.1016/j.ijhydene.2017.04.035.

Kieger, Stéphane, Gérard Delahay, Bernard Coq, and Bernard Neveu. (1999). Selective Catalytic Reduction of Nitric Oxide by Ammonia over Cu-FAU Catalysts in Oxygen-Rich Atmosphere. *Journal of Catalysis* 183 (2): 267–80. https://doi.org/10.1006/jcat.1999.2398.

Kovářík, Tomáš, Tomáš Křenek, David Rieger, Michal Pola, Jan Říha, Miloš Svoboda, Jan Beneš, Pavol Šutta, Petr Bělský, and Jaroslav Kadlec. (2017). Synthesis of Open-Cell Ceramic Foam Derived from Geopolymer Precursor via Replica Technique. *Materials Letters* 209: 497–500. https://doi.org/10.1016/j.matlet.2017.08.081.

Kumari, Snehlata, Biswajit Saha, Debdatta Guha, and Sonali Sengupta. (2021). Sequential Post-Synthetic Modification and Evaluation of Catalytic Activity of Hierarchically Porous Sulfated Geopolymer in the Oxidative Desulfurization of Dibenzothiophene. *Catalysis Surveys from Asia* 25 : 206-221. https://doi.org/10.1007/s10563-021-09328-6.

Landi, Elena, Valentina Medri, Elettra Papa, Jiri Dedecek, Petr Klein, Patricia Benito, and Angelo Vaccari. (2013). Alkali-Bonded Ceramics with Hierarchical Tailored Porosity. *Applied Clay Science* 73: 56–64. https://doi.org/10.1016/j.clay.2012.09.027.

Lemougna, Patrick N., A. Nzeukou, B. Aziwo, A. B. Tchamba, Kai-tuo Wang, U. Chinje Melo, and Xue-min Cui. (2020). Effect of Slag on the Improvement of Setting Time and Compressive Strength of Low Reactive Volcanic Ash Geopolymers Synthetized at Room Temperature. *Materials Chemistry and Physics* 239: 122077. https://doi.org/10.1016/j.matchemphys.2019.122077.

Li, Chun min, Yan He, Qing Tang, Kai tuo Wang, and Xue min Cui. (2016). Study of the Preparation of CdS on the Surface of Geopolymer Spheres and Photocatalyst Performance. *Materials Chemistry and Physics* 178: 204–10. https://doi.org/10.1016/j.matchemphys.2016.05.013.

Li, Wentao, and Yaolin Yi. (2020). Use of Carbide Slag from Acetylene Industry for Activation of Ground Granulated Blast-Furnace Slag. *Construction and Building Materials* 238: 117713. https://doi.org/10.1016/j.conbuildmat.2019.117713.

Lin, Hui, Hui Liu, Yue Li, and Xiangming Kong. (2021). Properties and Reaction Mechanism of Phosphoric Acid Activated Metakaolin Geopolymer at Varied Curing Temperatures. *Cement and Concrete Research* 144: 106425. https://doi.org/10.1016/j.cemconres.2021.106425.

Liu, Yuantao, Zhaohui Qin, and Bing Chen. (2020). Experimental Research on Magnesium Phosphate Cements Modified by Red Mud. *Construction and Building Materials* 231: 117131. https://doi.org/10.1016/j.conbuildmat.2019.117131.

Maiti, Moumita, Manas Sarkar, Soumen Maiti, Muhammad Akbar Malik, and Shilang Xu. (2020). Modification of Geopolymer with Size Controlled TiO_2 Nanoparticle for Enhanced Durability and Catalytic Dye Degradation under UV Light. *Journal of Cleaner Production* 255: 120183. https://doi.org/10.1016/j.jclepro.2020.120183.

Mardhiah, H. Haziratul, Hwai Chyuan Ong, H. H. Masjuki, Steven Lim, and H. V. Lee. (2017). A Review on Latest Developments and Future Prospects of Heterogeneous Catalyst in Biodiesel Production from Non-Edible Oils. *Renewable and Sustainable Energy Reviews* 67: 1225–36. https://doi.org/10.1016/j.rser.2016.09.036.

Matsuoka, Masaya, Masaaki Kitano, Masato Takeuchi, Koichiro Tsujimaru, Masakazu Anpo, and John M. Thomas. (2007). Photocatalysis for New Energy Production. Recent Advances in Photocatalytic Water Splitting Reactions for Hydrogen Production. *Catalysis Today* 122 (1–2): 51–61. https://doi.org/10.1016/j.cattod.2007.01.042.

Medpelli, Dinesh, Jung-min Seo, and Dong-Kyun Seo. (2014). Geopolymer with Hierarchically Meso-/Macroporous Structures from Reactive Emulsion Templating. *Journal of the American Ceramic Society* 97 (January). https://doi.org/10.1111/jace.12724.

Medri, Valentina, Elettra Papa, Jiri Dedecek, Hana Jirglova, Patricia Benito, Angelo Vaccari, and Elena Landi. (2013). Effect of Metallic Si Addition on Polymerization Degree of *in Situ* Foamed Alkali-Aluminosilicates. *Ceramics International* 39 (7): 7657–68. https://doi.org/10.1016/j.ceramint.2013.02.104.

Molaei Raisi, Elias, Javad Vaseghi Amiri, and Mohammad Reza Davoodi. (2018). Mechanical Performance of Self-Compacting Concrete Incorporating Rice Husk Ash. *Construction and Building Materials* 177: 148–57. https://doi.org/10.1016/j.conbuildmat.2018.05.053.

Nicolaides CP, Sincadu NP, Scurrell MS (2002). NAS (novel aluminosilicates) as catalysts for the aromatisation of propane: studies of zinc and gallium modified zeolite-based systems having various extents of XRD crystallinity. *Catalysis Today* 71: 429–435. https://doi.org/10.1016/S0920-5861(01)00470-9

Nie, Qingke, Wei Hu, Baoshan Huang, Xiang Shu, and Qiang He. (2019). Synergistic Utilization of Red Mud for Flue-Gas Desulfurization and Fly Ash-Based Geopolymer Preparation. *Journal of Hazardous Materials* 369: 503–11. https://doi.org/10.1016/j.jhazmat.2019.02.059.

Nimwinya, Ekkasit, Weerachai Arjharn, Suksun Horpibulsuk, Tanakorn Phoo-ngernkham, and Anurat Poowancum. (2016). A Sustainable Calcined Water Treatment Sludge and Rice Husk Ash Geopolymer. *Journal of Cleaner Production* 119: 128–34. https://doi.org/10.1016/j.jclepro.2016.01.060.

Novais, Rui M., João Carvalheiras, David M. Tobaldi, Maria P. Seabra, Robert C. Pullar, and João A Labrincha. (2019). Synthesis of Porous Biomass Fly Ash-Based Geopolymer Spheres for Efficient Removal of Methylene Blue from Wastewaters. *Journal of Cleaner Production* 207: 350–62. https://doi.org/10.1016/j.jclepro.2018.09.265.

O'Connor, Sean J., Kenneth J. D. MacKenzie, Mark E. Smith, and John V. Hanna. (2010). Ion Exchange in the Charge-Balancing Sites of Aluminosilicate Inorganic Polymers. *Journal of Materials Chemistry* 20 (45): 10234–40.

Oliver JM, López Nieto JM, Botella P, Mifsud A (2004). The effect of pH on structural and catalytic properties of MoVTeNbO catalysts. *Applied Catalysis A: General* 257: 67–76. https://doi.org/10.1016/S0926-860X(03)00632-X.

Papa, Elettra, Valentina Medri, Patricia Benito, Angelo Vaccari, Simone Bugani, Jakub Jaroszewicz, Wojciech Swieszkowski, and Elena Landi. (2015). Synthesis of Porous Hierarchical Geopolymer Monoliths by Ice-Templating. *Microporous and Mesoporous Materials* 215: 206–14. https://doi.org/10.1016/j.micromeso.2015.05.043.

Pathak, Uttarini, Snehlata Kumari, Anuj Kumar,, and Tamal Mandal. (2020). Process Parametric Optimization toward Augmentation of Silica Yield Using Taguchi Technique and Artificial Neural Network Approach. *Energy, Ecology and Environment* 5 : 294-312 https://doi.org/10.1007/s40974-020-00152-8.

Pelisser, Fernando, Adriano Michael Bernardin, Milton Domingos Michel, and Caroline Angulski da Luz. (2021). Compressive Strength, Modulus of Elasticity and Hardness of Geopolymeric Cement Synthetized from Non-Calcined Natural Kaolin. *Journal of Cleaner Production* 280: 124293. https://doi.org/10.1016/j.jclepro.2020.124293.

Prud'homme, E., E. Joussein, and S. Rossignol. (2015). Use of Silicon Carbide Sludge to Form Porous Alkali-Activated Materials for Insulating Application. *The European Physical Journal Special Topics* 224 (9): 1725–35. https://doi.org/10.1140/epjst/e2015-02494-7.

Prud'homme, E., P. Michaud, E. Joussein, C. Peyratout, A. Smith, and S. Rossignol. (2011). *In Situ* Inorganic Foams Prepared from Various Clays at Low Temperature. *Applied Clay Science* 51 (1): 15–22. https://doi.org/10.1016/j.clay.2010.10.016.

Rutkowska, Gabriela, Piotr Wichowski, Joanna Fronczyk, Małgorzata Franus, and Marek Chalecki. (2018). Use of Fly Ashes from Municipal Sewage Sludge Combustion in Production of Ash Concretes. *Construction and Building Materials* 188: 874–83. https://doi.org/10.1016/j.conbuildmat.2018.08.167.

Sanjayan, Jay G., Ali Nazari, Lei Chen, and Giang Hoang Nguyen. (2015). Physical and Mechanical Properties of Lightweight Aerated Geopolymer. *Construction and Building Materials* 79: 236–44. https://doi.org/10.1016/j.conbuildmat.2015.01.043.

Saputra, E., M. W. Nugraha, Z. Helwani, M. Olivia, and S. Wang. (2018). Synthesis of Geopolymer from Rice Husk Ash for Biodiesel Production of Calophyllum Inophyllum Seed Oil. *IOP Conference Series: Materials Science and Engineering* 345 (1). https://doi.org/10.1088/1757-899X/345/1/012019.

Sarkar, Chayan, Jayanta Kumar Basu, and Amar Nath Samanta. (2018). Synthesis of Mesoporous Geopolymeric Powder from LD Slag as Superior Adsorbent for Zinc (II) Removal. *Advanced Powder Technology* 29 (5): 1142–52. https://doi.org/https://doi.org/10.1016/j.apt.2018.02.005.

Saufi, Hamid, Marouane El Alouani, Saliha Alehyen, Mohammed El Achouri, Jilali Aride, and M'hamed Taibi. (2020). Photocatalytic Degradation of Methylene Blue from Aqueous Medium onto Perlite-Based Geopolymer. *International Journal of Chemical Engineering* 2020 : 9498349. https://doi.org/10.1155/2020/9498349.

Sayed, Maha R., Mostafa R. Abukhadra, Sayed Abdelkader Ahmed, Mohamed Shaban, Umer Javed, Mohamed A. Betiha, Jae Jin Shim, and Abdelrahman M. Rabie. (2020). Synthesis of Advanced MgAl-LDH Based Geopolymer as a Potential Catalyst in the Conversion of Waste Sunflower Oil into Biodiesel: Response Surface Studies. *Fuel* 282 (August): 118865. https://doi.org/10.1016/j.fuel.2020.118865.

Sazama, Petr, Oleg Bortnovsky, Jiří Dědeček, Zdenka Tvarůžková, and Zdeněk Sobalík. (2011). Geopolymer Based Catalysts-New Group of Catalytic Materials. *Catalysis Today* 164 (1): 92–99. https://doi.org/10.1016/j.cattod.2010.09.008.

Schwieger, Wilhelm. (2016). Hierarchy Concepts: Classification and Preparation Strategies for Zeolite Containing Materials with Hierarchical Porosity. *Chemical Society Reviews* 45 (12): 3353–76. https://doi.org/10.1039/C5CS00599J.

Seelam, Prem Kumar, Harisankar Sreenivasan, Satu Ojala, Satu Pitkäaho, Tiina Laitinen, He Niu, Riitta L. Keiski, and Mirja Illikainen. (2021). Modified Geopolymers as Promising Catalyst Supports for Abatement of Dichloromethane. *Journal of Cleaner Production* 280: 124584. https://doi.org/10.1016/j.jclepro.2020.124584.

Sellami, Mouna, Maud Barre, and Mohamed Toumi. (2019). Synthesis, Thermal Properties and Electrical Conductivity of Phosphoric Acid-Based Geopolymer with Metakaolin. *Applied Clay Science* 180: 105192. https://doi.org/10.1016/j.clay.2019.105192.

Serrano, D. P., J. M. Escola, and P. Pizarro. (2013). Synthesis Strategies in the Search for Hierarchical Zeolites. *Chemical Society Reviews* 42 (9): 4004–35. https://doi.org/10.1039/C2CS35330J.

Sharma, Sudhanshu, Dinesh Medpelli, Shaojiang Chen, and Dong-Kyun Seo. (2015). Calcium-Modified Hierarchically Porous Aluminosilicate Geopolymer as a Highly Efficient Regenerable Catalyst for Biodiesel Production. *RSC Advances* 5 (July). https://doi.org/10.1039/C5RA01823D.

Strini, Alberto, Giuseppina Roviello, Laura Ricciotti, Claudio Ferone, Francesco Messina, Luca Schiavi, Davide Corsaro, and Raffaele Cioffi. (2016). TiO_2-Based Photocatalytic Geopolymers for Nitric Oxide Degradation. *Materials* 9 (7): 1–13. https://doi.org/10.3390/ma9070513.

Subaer, Abdul Haris, A Irhamsyah, Nur Akifah, and Nur Syamsih Amalia. (2019). Physico-Mechanical Properties of Geopolymer Based on Laterite Deposit Sidrap, South Sulawesi. *Journal of Physics: Conference Series* 1244 (1): 12037. https://doi.org/10.1088/1742-6596/1244/1/012037.

Tang, Qing, Yuan-yuan Ge, Kai-tuo Wang, Yan He, and Xue-min Cui. (2015). Preparation and Characterization of Porous Metakaolin-Based Inorganic Polymer Spheres as an Adsorbent. *Materials & Design* 88: 1244–49. https://doi.org/10.1016/j.matdes.2015.09.126.

Tang, Qing, Kaituo Wang, Jianfeng Su, Yingqian Shen, Sijie Yang, Yuanyuan Ge, and Xuemin Cui. (2019). Facile Fabrication of Inorganic Polymer Microspheres as Adsorbents for Removing Heavy Metal Ions. *Materials Research Bulletin* 113: 202–8. https://doi.org/10.1016/j.materresbull.2019.02.009.

Wan, Hengyu, Yan He, Qiaoqiao Su, Leping Liu, and Xuemin Cui. (2022). Slag-Based Geopolymer Microspheres as a Support for CO_2 Methanation. *Fuel* 319: 123627. https://doi.org/10.1016/j.fuel.2022.123627.

Yan, Shu, Peigang He, Dechang Jia, Zhihua Yang, Xiaoming Duan, Shengjin Wang, and Yu Zhou. (2015). *In Situ* Fabrication and Characterization of Graphene/Geopolymer Composites. *Ceramics International* 41 (9, Part A): 11242–50. https://doi.org/10.1016/j.ceramint.2015.05.075.

Yang, Zhaotong, Ryan Mocadlo, Mengxuan Zhao, Richard D Sisson, Mingjiang Tao, and Jianyu Liang. (2019). Preparation of a Geopolymer from Red Mud Slurry and Class F Fly Ash and Its Behavior at Elevated Temperatures. *Construction and Building Materials* 221: 308–17. https://doi.org/10.1016/j.conbuildmat.2019.06.034.

Zhang, Baifa, Haozhe Guo, Liangliang Deng, Wenxiao Fan, Ting Yu, and Qiang Wang. (2020). Undehydrated Kaolinite as Materials for the Preparation of Geopolymer through Phosphoric Acid-Activation. *Applied Clay Science* 199: 105887. https://doi.org/10.1016/j.clay.2020.105887.

Zhang, Baifa, Haozhe Guo, Peng Yuan, Yun Li, Qiang Wang, Liangliang Deng, and Dong Liu. (2020). Geopolymerization of Halloysite via Alkali-Activation: Dependence of Microstructures on Precalcination. *Applied Clay Science* 185: 105375. https://doi.org/10.1016/j.clay.2019.105375.

Zhang, Weiqing, Chaowei Dong, Peng Huang, Qiang Sun, Meng Li, and Jun Chai. (2020). Experimental Study on the Characteristics of Activated Coal Gangue and Coal Gangue-Based Geopolymer. *Energies* . https://doi.org/10.3390/en13102504.

Zhang, Yao Jun, and Qian Chai. (2014). Alkali-Activated Blast Furnace Slag-Based Nanomaterial as a Novel Catalyst for Synthesis of Hydrogen Fuel. *Fuel* 115: 84–87. https://doi.org/10.1016/j.fuel.2013.06.051.

Zhang, Yao Jun, Pan Yang He, and Hao Chen. (2018). A Novel CdO/Graphene Alkali-Activated Steel Slag Nanocomposite for Photocatalytic Degradation of Dye Wastewater. *Ferroelectrics* 522 (1): 1–8. https://doi.org/10.1080/00150193.2017.1391587.

Zhang, Yao Jun, Pan Yang He, Meng Yang Yang, and Le Kang. (2017). A New Graphene Bottom Ash Geopolymeric Composite for Photocatalytic H_2 Production and Degradation of Dyeing Wastewater. *International Journal of Hydrogen Energy* 42 (32): 20589–98. https://doi.org/10.1016/j.ijhydene.2017.06.156.

Zhang, Yao Jun, Pan Yang He, Yi Xin Zhang, and Hao Chen. (2018). A Novel Electroconductive Graphene/Fly Ash-Based Geopolymer Composite and Its Photocatalytic Performance. *Chemical Engineering Journal* 334 (13): 2459–66. https://doi.org/10.1016/j.cej.2017.11.171.

Zhang, Yao Jun, Li Cai Liu, Lu Lu Ni, and Bing Li Wang. (2013). A Facile and Low-Cost Synthesis of Granulated Blast Furnace Slag-Based Cementitious Material Coupled with Fe_2O_3 Catalyst for Treatment of Dye Wastewater. *Applied Catalysis B: Environmental* 138–139 (x): 9–16. https://doi.org/10.1016/j.apcatb.2013.02.025.

Zhang, Yao Jun, Meng Yang Yang, Li Zhang, Ke Zhang, and Le Kang. (2016). A New Graphene/Geopolymer Nanocomposite for Degradation of Dye Wastewater. *Integrated Ferroelectrics* 171 (1): 38–45. https://doi.org/10.1080/10584587.2016.1171178.

Zhang, Yao Jun, Ke Zhang, Le Kang, and Li Zhang. (2017). A New In_2O_3 and NiO Co-Loaded Fly Ash-Based Nanostructural Geopolymer for Photocatalytic H_2 Evolution. *Integrated Ferroelectrics* 182 (1): 1–9. https://doi.org/10.1080/10584587.2017.1352376.

Zhang, Yao Jun, Li Zhang, Le Kang, Meng Yang Yang, and Ke Zhang. (2017). A New $CaWO_4$/Alkali-Activated Blast Furnace Slag-Based Cementitious Composite for Production of Hydrogen. *International Journal of Hydrogen Energy* 42 (6): 3690–97. https://doi.org/10.1016/j.ijhydene.2016.07.173.

Zhang, Yao Jun, Yi Xin Zhang, and Meng Yang Yang. (2018). Synthesis of Environment-Friendly Graphene Reinforced Slag-Based Nanocomposite and Performance of Photocatalytic H_2 Generation. *Ferroelectrics* 522 (1): 36–44. https://doi.org/10.1080/00150193.2017.1391609.

Zhang, Yaojun, and Licai Liu. (2013). Fly Ash-Based Geopolymer as a Novel Photocatalyst for Degradation of Dye from Wastewater. *Particuology* 11 (3): 353–58. https://doi.org/10.1016/j.partic.2012.10.007.

Zhu, Yifan, Shanguo Ji, Wenxu Liang, Changyu Li, Yongxin Nie, Jing Dong, Weijie Shi, and Shiyun Ai. (2022). A Low-Cost and Eco-Friendly Powder Catalyst: Iron and Copper Nanoparticles Supported on Biochar/Geopolymer for Activating Potassium Peroxymonosulfate to Degrade Naphthalene in Water and Soil. *Chemosphere* 303: 135185. https://doi.org/10.1016/j.chemosphere.2022.135185.

Chapter 2

Study of the Adsorption of Cu(II), Mn(II), Pb(II), and Zn(II) Ions on Geopolymers Obtained from Ashes from Biomass Combustion

E. Sitarz-Palczak[1,*], D. Galas[2] and J. Kalembkiewicz[1]

[1]Department of Inorganic and Analytical Chemistry, Faculty of Chemistry, Rzeszow University of Technology, Rzeszów, Poland

[2]Regional Directorate for Environmental Protection, Rzeszów, Poland

Abstract

Several industries produce wastewater containing a wide range of pollutants. This includes organic impurities such as dyes, petroleum products, pharmaceuticals, and inorganic pollutants, which mainly contain heavy metals. It is widely believed that both dyes and heavy metals are among the most dangerous contaminants of water systems due to their mutagenic, carcinogenic, and teratogenic effects. Efficient wastewater treatment using conventional methods is often limited due to the low selectivity and high costs of such methods. An ecological, economical, and efficient alternative to conventional methods is the adsorption process on the geopolymers obtained from the by-products of combustion, i. e. the ash from biomass combustion.

The aim of the work was to investigate the adsorption of Cu(II), Mn(II), Pb(II), and Zn(II) ions to geopolymers obtained from biomass combustion ash. The influence of the starting metal concentration in the aqueous solution and the adsorption duration on the adsorption efficiency of the ions studied was discussed. The Langmuir and Freundlich models

* Corresponding Author's Email: epalczak@prz.edu.pl.

In: Geopolymers
Editor: 979-8-88697-197-2
ISBN: Joseph E. Glenn

were used to describe the adsorption equilibrium of Cu(II), Mn(II), Pb(II), and Zn(II) ions from aqueous solutions on geopolymers obtained from the ash of biomass combustion.

Analysis of the results showed that the adsorption efficiency of the tested ions increases over time in the following ranges: Cu(II) 50–95%, Mn(II) 41–92%, Pb(II) 33–89%, Zn(II) 58–90%. Experimental sorption content values are 133 mg/g for Cu(II), 105 mg/g for Mn(II), 348 mg/g for Pb(II), and 132 mg/g for Zn(II). The adsorption of the investigated ions to geopolymers extracted from the ash of biomass combustion is carried out according to a pseudo-second order model. Analysis of the FTIR spectra of geopolymers after adsorption of the tested metal ions indicates a proportion of functional groups, e.g., O-H, N-H, and C-O, by adsorption. Comparison of all tested geopolymers that have adsorbed Cu(II), Mn(II), Pb(II), and Zn(II) ions showed that they have the smallest size for the geopolymer after adsorption of Pb(II) ions and the largest for the geopolymer after adsorption of Zn(II) ions.

In summary, geopolymers from biomass combustion by adsorption of Cu(II), Mn(II), Pb(II), and Zn(II) ions from aqueous solutions can be used. The geopolymers used have good sorption properties compared to the heavy metal ions studied, indicating that they can be used as adsorption agents for industrial and environmental pollutants. In addition, the proposed application is an example of the use of geopolymers as recycled products of combustion by-products.

Keywords: geopolymer, biomass, adsorption, heavy metals

Introduction

Biomass is an organic-inorganic solid product produced by natural and anthropogenic (technological) processes, which includes natural components derived from the cultivation of terrestrial and aquatic vegetation through photosynthesis or produced by digestion of food by animals and humans, and technological products obtained by processing the above natural components [1]. Biomass sources include specially grown energy crops, agricultural waste, forest residues, and the organic fraction of municipal waste. Biomass has a wide variety of physical properties that have a significant impact on the rate of biomass conversion processes [2]. Unprocessed biomass is only a raw material that, after processing, becomes a proper fuel. Biomass processing aims to adapt its physical properties to plant requirements. Untreated biomass, especially wood, contains large amounts of water and low calorific value [3]. Therefore, biomass is subjected to various conversion processes using physical, chemical, and biochemical methods. These processes result in biofuels that are used for energy purposes. The most important biomass processing processes include drying, shredding (chips), gasification, pyrolysis, conditioning (briquette), oil esterification, alcoholic and methane fermentation [4]. Many biomass feedstocks, especially grasses, straw, and sawdust, have a very low bulk density (usually less than 150 kg/m^3), which requires high storage, transportation, and handling costs. Therefore, palletization, which involves densifying biomass to more than 600 kg/m^3, is often used to make biomass competitive with other energy sources [2].

Biomass fuels are the main source of energy for more than 50% of the world population and account for about 14% of total global energy consumption. On the other hand, biomass corresponds to the most popular form of renewable energy, which is becoming increasingly interesting. It is believed to play an important role in mitigating global warming and securing fuel supply. Over the past decade, interest in biomass has resurfaced internationally because it is the most promising alternative to fossil fuels. In addition, biomass is readily available in many countries around the world, especially in developing countries. This renewable energy resource can be sustainable if it is properly managed and provides a local supply of fuel at competitive prices. The use of biomass as a fuel also provides environmental benefits. When properly operated and controlled, burning biomass produces significantly less nitrogen oxides and sulfur dioxide than solid fossil fuels, such as coal and lignite. Biomass is an excellent source of energy; its conversion to energy; occurs through pyrolysis, direct combustion, gasification, liquefaction, anaerobic digestion, alcoholic digestion, and transesterification. The choice depends on the physicochemical properties of the biomass and the amount of energy required for the process [4].

The process of energy and heat production through biomass combustion can be carried out in various ways, namely direct combustion in boilers, co-combustion with conventional energy carriers, and combustion of products derived from biomass processing. Regardless of the combustion method, the process itself generates large amounts of fly ash in the environment. Traditionally, these ashes are disposed of by depositing them in special landfills. However, with the increase in cost, purchase, and development of new landfill sites, fly ash is a serious problem for the power generation industry. Fly ash generated from biomass consists mainly of an inorganic fraction and a small amount of organic fraction (unburned carbon). The proportions of these fractions depend on various parameters and conditions of combustion processes, including the type of biomass and the installation used

[5]. A characteristic feature of fly ashes is the presence of elements such as calcium, magnesium, potassium, phosphorus, aluminum, and sulphur in their composition. Furthermore, they may also contain heavy metals, such as manganese, barium, boron, copper, nickel, chromium, cadmium, cobalt, molybdenum, and zinc, which constitute about 0.1-0.3% of the total mass of ashes. The reaction of fly ash is alkaline and ranges between 9.3 and 13.9 pH, it depends mainly on the type of plant biomass used [6]. Utilizing the ash from biomass combustion is part of a sustainable energy generation process, while storing it can be interpreted as a loss of valuable nutrients contained in the ash. The search for utilization must address very different types of ash. The composition of the fuel and the type of plant are the main factors that affect the quality of the ash. Changes in the inorganic fraction of fuels are directly reflected in the composition of the ash. Different types of installations produce many types of ash, each with different morphological characteristics, composition, and leaching behavior. Biomass ash should be recycled whenever possible, but the quality of the ash must be high enough to prevent contamination by spreading ash. Ash recycling occurs in many countries, but the total amounts recycled are very small compared to the total biomass ash production [5].

The ashes obtained from biomass combustion can be used directly as a fertilizer or soil improver, as well as a raw material for mineral fertilizers. The ashes can be returned to the soil, but not necessarily to the same place where the biomass was taken. The use of biomass ash in general-purpose fertilizers is limited. The three primary elements in fertilizers are N, P, and K. Biomass ash can only be a significant source of potassium because thermal process ash is almost free of nitrogen, and phosphorus is present in a form that has very poor solubility under soil conditions. In practice, biomass combustion ash is not attractive as a feedstock because it has a low nutrient ratio compared to the pollutants present [1].

The use of ash as a fuel in combustion processes does not completely solve the problem of finding a sustainable way to use biomass ash. Incineration produces low-emission combustion ash that is virtually identical to biomass combustion ash. There is no market for ash fuels, but preliminary estimates indicate that this is the most economical way to use fly ash with a high unburned carbon content. In practice, this option is limited to carbon-rich fly ash from fluidized bed gasification [7, 8].

Fly ash from biomass combustion, which exhibits alkalizing properties and the presence of reactive calcium oxide causes, can be used as reactants in the process of higienizing sewage sludge. The addition of ash to the sewage sludge causes the occurrence of various exothermic reactions in which the sludge undergoes disinfection, sterilization, and pasteurization processes. As the amount of ash added to the sewage sludge increases, the pH of the resulting mixture increases, causing pathogenic organisms to die. Hygienization of wastewater sludge with biomass ash allows the reduction of pathogenic bacteria in the range of 93-100% [7]. Ash is also used in underground mining as a component of materials used to fill underground works, to liquidate and fill old mine bogs, and to build explosion-proof walls [8].

Biomass combustion ash is used in the food industry as a filler that improves mechanical properties, reduces flammability, and costs of plastic production. The process of filling plastics with ash is carried out in such a way that plastics retain their characteristic functional properties [8].

Another possible application of the ashes from biomass combustion is their use as a building material or raw material for construction products. However, these ashes must be subject to the same technical criteria and environmental regulations as any other material. Fly ash from biomass combustion is one of the most difficult materials to use. Direct use as a loose building material to replace sand or gravel is almost impossible, because it is a fine powder. The use of cement or concrete as a component is a more likely option. One way to use biomass fly ash is as a filler in cement mixes or in specialty mortars. A frequently mentioned option is the production of lightweight aggregates such as Lytag. The use of biomass combustion ash for concrete production has been described in papers [9, 10], among others.

Research on the possibilities of nonagricultural utilization of biomass combustion ash has been conducted for a number of years. Unconventional methods include the processing of biomass ash into adsorbents, zeolites, and geopolymers. The good adsorption properties of ashes are mainly due to the developed surface area and high ion exchange capacity. In addition, the ashes have a strongly alkaline reaction and a negatively charged surface, so that positive metal ions can be adsorbed by them. The use of ash from biomass combustion as an adsorbent is indicated to be a cheaper solution compared to activated carbon. Literature reports on this subject are mainly related to the sorption of metals: As, Au, Cd, Co, Cr, Cu, Hg, Mn, Ni, Pb, and Zn, and nonmetals - boron, fluorine and chemical compounds such as NH_3, NOX, SOX, phenols, toluene, polychlorinated aromatic hydrocarbons; dyes [11].

One of the promising possibilities for the processing of combustion by-products such as ashes is the synthesis of zeolites and geopolymers based on them. The properties of ashes, especially the significant content of aluminum and silicon, create the possibility of producing synthetic zeolites, which are an attractive material for use in environmental protection [11]. The data of the literature show [12] that using a properly selected methodology, zeolites such as analcime, chabazite, cancrinite, gmelinite, NaP1, ZSM-5, ZMS-28, Na-X, Na-Y, phillipsite, and sodalite can be obtained from ashes. Previous studies indicate that zeolites obtained from ashes can be used to adsorption of SO_2 and NO_x, remove heavy metals and ammonium ions from wastewater, and remediate groundwater by building reactive barriers [13].

The use of combustion by-products to obtain geopolymers is a pro-ecological approach because it allows for practical utilization of waste. Up to now, research has focused mainly on the use of industrial ash from coal combustion [14], although in recent years work has been carried out on the possibility of using ash from biomass combustion and cofiring of biomass with coal for the synthesis of geopolymers.

The choice of ash for geopolymer synthesis is determined by a number of factors that affect the ease and speed of production, as well as the final characteristics of the finished products. The most important characteristics of the starting material include:

- morphology of ash particles: regular structure of grains positively affects workability of the mixture, reduces the number of liquid substances, and improves rheological properties;
- content of SiO_2 and Al_2O_3 – a high content is necessary to obtain geopolymer materials;
- loss of ignition; high content results in lower silica content and increases the water content of the ash;
- a high content of iron compounds has a negative effect on the ash particle size distribution, as these compounds usually accumulate on the grain surfaces and thus obstruct the access of the liquid phase to the glassy phase;
- amount of calcium oxide affects the setting rate, and too high a content results in insufficient dissolution of active ash components in the geopolymer mass [15, 16].

The ashes from biomass combustion consist mainly of amorphous silica and alumina, which makes them suitable materials for the production of geopolymers [17, 18]. However, the great diversity of this group of industrial precipitates makes it necessary to consider each type of ash individually.

Bajere et al. [19] obtained geopolymers based on ash from wood combustion, as well as barley. NaOH was used as an alkaline activator. The materials obtained had compressive strengths of 2.8 MPa and 3.2 MPa using wood and barley ash, respectively. Rajamana et al. [20] studied the effect of the composition of the wood ash-based geopolymer mix (different content of the alkali activator - water glass + NaOH) on the quality of the product obtained. The compressive strength of the geopolymers obtained ranged from 8.2 to 11.7 MPa, the highest recorded when the highest NaOH concentration was used. Candamano et al. [21] investigated the possibility of obtaining metakaolin based geopolymers with the addition of ash from forest biomass combustion (10 - 30%), used a mixture of sodium silicate and sodium hydroxide as an alkaline activator. The syntheses were carried out at an optimum Si/Al ratio of 2. All the materials obtained had compressive strengths above 30 MPa.

Rice husk combustion ash shows a high silica content, which makes it used for cement production. Kaur et al. [22] obtained geopolymers based on rice husk ash, they used NaOH at different concentrations as an alkaline activator. The compressive strength of the geopolymers depended on the proportion of components used during synthesis and was 9.12 - 28.79 MPa, 11.46 - 33.52 MPa, 20.19 - 36.49 MPa and 28.56 - 39.95 MPa after 3, 7, 14 and 28 days, respectively. The geopolymer with an ash/activator ratio of 0.7 had the highest strength, while the geopolymer with a ratio of 0.5 had the lowest strength. Increasing the concentration of alkali activators leads to a higher strength of the geopolymer because of the denser microstructure and homogeneous matrix. Sturm et al. [23] prepared a geopolymer based on rice husk combustion ash using sodium aluminate as an alkaline activator. The molar ratio of Na_2O:Al_2O_3 in the geopolymer mixture was 1, while the ratio of SiO_2:Al_2O_3 was 3.48. The geopolymer obtained had strengths of 29.8, 32.7, and 30.1 MPa after 1, 3, and 7 days. On the other hand, Detphan and Chindaprasirt [24] investigated the possibility of using a mixture of coal ash cofiring with rice husks (in proportions 100/0, 80/20, 60/40, 40/60) for the synthesis of geopolymers, and used sodium hydroxide and sodium silicate as alkaline activator. The materials obtained exhibited a high compressive strength, ranging from 12.5 - 56 MPa, depending on the synthesis conditions. Samadhi et al. [25] prepared inorganic aluminosilicate polymers using waste from the combustion of rice husks and coconut husks, Na_2SiO_3/KOH was used as an alkaline activator. The compressive strengths of the geopolymers obtained varied and were 1.4 MPa (based on rice husk ash) and 3.1 MPa (based on coconut husk ash).

He et al. [26] investigated the possibility of obtaining a geopolymer composite using rice husk ash and red sludge (from aluminum production).

NaOH was used as an alkaline activator. The resulting material had a compressive strength comparable to most Portland cements of about 21 MPa. However, the long curing time, variability in composition, and possible side reactions hinder its practical application on a mass scale.

The unique properties of inorganic aluminosilicate polymers make them increasingly widely used [14, 27]. The main proposed direction of their use is the building materials industry. Geopolymer cement is an innovative material and is an alternative to conventional Portland cement [28]. Geopolymer materials because of their properties, i.e., high mechanical strength, weather resistance, and fire resistance, are also used in specific cases. As examples of applications of geopolymers [14, 15, 27] the following are indicated: emergency runway repairs, fire insulation in aviation, sandwich panels, foamed geopolymer panels for thermal insulation, self-hardening masses in foundry, fire composites. Geopolymer materials can also be used as reinforcement and sealing material used in drilling, mining, and hydraulic engineering [29].

Intensive work has been carried out in the world for several years on the application of geopolymers as a protective coating of materials [27]. Zhang et al. [30] developed a reflective thermal insulation coating using a metakaolin-based geopolymer as the basic film-forming material. Among others, the coating is characterized by simple atomization, high durability and stain resistance, a reflectance more than 90%, and high water retention, suggesting its potential application in energy-efficient construction. Geopolymer coatings can also be used as surface protection for building structures to extend their service life. Wiyono et al. [31] found that the application of a geopolymer protective coating to concrete results in a reduction in the chloride ion transfer coefficient while improving resistance to 10% sulfuric (VI) acid solution. Geopolymer coatings can be used to protect against seawater [32]. A very important application area of geopolymer-based coatings is the protection of steel against corrosion. Its application results in the improvement of properties related to the heat resistance and fire resistance of steel. However, proper coating selection is very important, especially due to thermal expansion of the base material and coating [33].

Geopolymers are also predicted to be used as catalyst materials. Sazama et al. [34] developed a new type of heterogeneous redox catalysts based on geopolymer materials. Transition-metal ions such as: Fe(II), Co(II), and Cu(II) can be incorporated into geopolymer networks and form active centers for catalytic reactions. The authors investigated the applicability of geopolymer catalysts for NO_x reduction and hydrocarbon oxidation. The materials tested showed high conversion rates of nitric oxide to nitrogen and decane to carbon dioxide [34]. Geopolymer materials are also predicted to be used, among others, in the creation of electronic composites with carbon nanotubes, photoactive composites with nanoparticles, or new chromatographic materials [27].

One of the new directions of geopolymer applications is their application as a carrier material for the disposal of toxic waste, including radioactive substances. In the geopolymerization process, hazardous wastes are disposed of simultaneously through a physical process (encapsulation of elements in the material) and by embedding them into the resulting network through chemical reactions. The geopolymer matrix is a good material for heavy metal immobilization due to its low permeability and resistance to acid and chloride ions. Ash-based geopolymers show a higher heavy metal-ion immobilization ability than Portland cement or ash alone [14, 15, 35]. Studies conducted by Zhang et al. [36] show that Pb(II) ions can be effectively immobilized with fly ash-based geopolymer by binding Pb(II) in the form of insoluble silicate (Pb_3SiO_5). El-Eswed et al. [37] also showed that heavy metals Pb(II), Cu(II), Cd(II), and Cr(III) can be immobilized using kaolin-based geopolymer. There are three possible mechanisms for the immobilization of heavy metals in the geopolymer matrix:

- ion exchange of Na^+, K^+ cations by heavy metal ions, which balances the negative charge of fragments of geopolymer chains containing aluminum (electrostatic interaction),
- formation of covalent bonds between heavy metals and the aluminosilicate chain,
- formation of hydroxides, carbonates, silicates, and heavy metal aluminates [38].

The physicochemical properties, porous structure, and negatively charged surface make geopolymers suitable for use as adsorbents. Today, there is ongoing research on the possibilities of their application for the adsorption of environmental pollutants such as heavy metals, e.g., Cd(II), Ni(II), and Pb(II), and dyes, e.g., methylene blue, congo red, methyl violet [39]. The powder and granular, monolithic, and spherical forms of the geopolymer are used for sorption. The process is influenced by several factors, such as geopolymer composition, adsorbent dosage, pH, temperature, contact time, and initial contaminant concentrations [40]. Reports in the literature [41-50] indicate that selected geopolymers exhibit good sorption properties. Their additional advantage is their low price, which is compared to conventional materials used as sorbents.

The present study deals with the application of geopolymers obtained from biomass combustion ashes for the adsorption of selected heavy metal ions from aqueous solutions.

Material and Methods

Material

The geopolymer obtained from the ash, which is a by-product of the combustion of deciduous wood chips, was used as a research material in this study. Biomass combustion ash is a source of reactive silica and alumina, which act as binders for the geopolymer matrix. It was used as a solid mixture composed of equal parts by weight of quartz sand and ash. A mixture of sodium silicate (oxide content SiO_2 + Na_2O - not less than 39% by weight; molar modulus SiO_2/Na_2O = 2.4-2.6) and solid sodium hydroxide (purity 97% by weight) was used as an alkaline activator. To obtain the geopolymer paste, the mixture of solid components was mixed with an alkaline activator in a 50:50 ratio of the aqueous phase [cm^3] to the solid phase [g]. After obtaining a homogeneous mixture, the geopolymer mortars were transferred to cylindrical molds (14 mm × 25 mm). The samples were removed from the molds after 24 h and then cured at room temperature (~20°C) and humidity (~68%) for 28 days. A detailed characterization of the starting raw materials and geopolymers obtained, as well as their synthesis conditions, is included in the work [51, 52].

Methods

Surface Morphology and Chemical Composition Analysis

Scanning electron microscopy (SEM) was used to characterize the surface of the geopolymers, using a scanning electron microscope equipped with a microarea chemical analysis (EDS) attachment, model S-3400 N (HITACHI, Japan). Preparations were made by placing the ash on carbon tape with gold sputtering as a necessary condition to avoid artifacts resulting from electrical charging of the surface of nonconductive materials during scanning by the electron beam [53].

Structural Parameters

A standard method for the analysis of adsorption isotherms in low-temperature nitrogen sorption based on the linear form of the BET isotherm equation (BET method) was used to determine the specific surface area, average pore volume, and average pore diameter. Nitrogen sorption isotherms for the subjects were determined at 77K (-196°C) using a specific surface area analyzer (model ASAP 2420, Micromeritics, USA). Before measurements, the samples were degassed at 120°C under a pressure of 0.001 Pa. From the determined adsorption capacity corresponding to the formation of the monolayer, the specific surface area of the studied samples was determined.

Spectroscopic Analysis

Infrared absorption spectra were recorded in the fundamental range of 4000 - 400 cm^{-1} with a resolution of 2 cm^{-1} using an FTIR spectrometer (Alpha model, BRUKER, Germany). Powder preparations for analysis were prepared by mixing the investigated samples with spectrally pure KBr. The mixture prepared in this way was pressed under vacuum at a pressure of 10 MPa, obtaining a preparation in the form of a lozenge.

Methodology of Investigations

Studies on adsorption of Cu(II), Mn(II), Pb(II), and Zn(II) ions on geopolymers obtained from ashes of biomass combustion were carried out based on sorption and desorption experiments.

Sorption Experiment

To determine the equilibrium adsorption isotherms of the studied metal ions, solutions of the corresponding salts $Cu(NO_3)_2$, $Mn(NO_3)_2$, $Zn(NO_3)_2$ and $Pb(NO_3)_2$ were prepared with the following concentrations: 0.2; 0.1; 0.05; 0.02; 0.01; 0.005; 0.0025 $mol{\cdot}dm^{-3}$. All sorption experiments were conducted in single-component systems. They consisted of shaking geopolymer samples with salt solutions of the studied metal ions, in which the ratio of the mass of the solid sample (1.0000 g) to the volume of the solution (100 cm^3) was 1:1. The experiments were performed at room temperature without correction of pH of the initial salt solutions, i.e., at the pH that was obtained after the salt of the studied metal was dissolved in aqueous solution. Shaking was carried out for 24 h at room temperature at a speed of 300 rpm using a laboratory shaker model Vibramax 100 (Heidolph Instruments, Germany). After 24 h, a 5 cm^3 sample volume was collected and the sample was filtered through a syringe filter (LGC Genomics, United Kingdom). All sorption experiments were performed in duplicate to minimize the risk of errors. The precipitate in each case was dried at room temperature and ground in an agate mortar. In order to identify the surface functional groups present on the surface of the geopolymers, we studied the solid residue obtained by FTIR. The concentrations of the studied metals in the obtained solutions were determined by atomic absorption spectroscopy under standard conditions summarized in Table 1.

The ionic content in the solid phase was calculated according to Equation (1):

$$q_e = \frac{C_o - C_e}{m} V \quad (1)$$

where: q_e - solid phase ion content [$mg \cdot g^{-1}$], C_0 - initial concentration of ions in solution [$mg \cdot dm^{-3}$], C_e - equilibrium concentration of ions in solution [$mg \cdot dm^{-3}$], V - volume of solution [cm^3], m - sorbent mass [g].

The adsorption coefficient (A) expressed in % was calculated using the relation (2):

$$A = \frac{C_o - C_e}{Co} 100\% \quad (2)$$

where: A - adsorption coefficient [%], C_0 - initial concentration of ions in solution [$mg \cdot dm^{-3}$], C_e - equilibrium concentration of ions in solution [$mg \cdot dm^{-3}$].

To determine the effect of time on the sorption process of Cu(II), Mn(II), Pb(II) and Zn(II) ions, solutions of the respective $Cu((NO_3)_2$, $Mn(NO_3)_2$, $Zn(NO_3)_2$ and $Pb(NO_3)_2$ salts were prepared at a concentration of 0.05 $mol \cdot dm^{-3}$. Then 1 g (± 0.0001) of geopolymers were weighed and 100 cm^3 of each of the prepared metal salt solutions were added. The resulting suspensions were shaken at room temperature at a speed of 300 rpm using a laboratory shaker, Vibramax model 100 (Heidolph Instruments, Germany). Samples of 5 cm^3 were collected after 0, 5, 15, 30, 60, 180, 300 1440 min;

after collection, the sample was filtered through a syringe filter (LGC Genomics, United Kingdom). The concentrations of the test metals in the solutions obtained were determined using the FAAS method under the standard conditions as subjected in Table 1.

Table 1. Conditions for the determination of the tested heavy metals using the FAAS method

Parameter	Cd	Cu	Ni	Pb	Zn
Wavelength [nm]	228.8	324.8	341.5	217.0	213.9
Sensitivity [mg/dm^3]	0.028	0.45	0.40	0.19	0.084
Slot width [nm]	0.7	0.7	0.2	0.7	0.7
Concentration of standard solutions [mg/cm^3]	1.5 1.0 2.0	2.0 4.0 5.0	5.0 10.0 20.0	9.0 20.0	1.0 3.0 5.0 6.0
Gas flow rate [dm^3/min]	0.8 – 1.0				

Desorption Experiment

The stability of the bond between the sorbate and the sorbent was determined by the desorption rate. Experiments of the desorption process were carried out using 1 $mol \cdot dm^{-3}$ ammonium acetate solution at pH = 7, which allows the removal of metal ions from the ion exchange positions. The desorption experiments were conducted under conditions identical to those of the sorption experiments. The precipitate in each case was dried at room temperature and ground in an agate mortar. In order to identify the surface functional groups present on the surface of the geopolymers, we studied the solid residue obtained by FTIR. The concentrations of the studied metals in the solutions obtained were determined by the FAAS method under the standard conditions as set out in Table 1.

Results and Discussion

Studies on Adsorption of Cu(II), Mn(II), Pb(II), and Zn(II) Ions on Geopolymers Obtained from Biomass Combustion Ashes

Effect of Initial Metal Concentration in Aqueous Solution on Adsorption Efficiency

The efficiency of the copper(II), manganese(II), lead(II), and zinc(II) adsorption process using geopolymers based on biomass combustion ashes was studied, and the influence of the initial metal concentration on the adsorption process was studied. Studies were carried out for the following concentration ranges of each metal: 159 - 12710 (Cu), 137 - 10988 (Mn), 518 - 41441 (Pb) and 164 - 13075 [mg·dm^{-3}] (Zn). The results of the analysis are presented in Table 2.

Table 2. Influence of the initial concentration of the studied metal ions on the adsorption by geopolymers
(m = 1.00 g, V = 50 cm^3, pH = 4, t = 24 h, T = 20°C)

Metal ion	Initial concentration [mg·dm^{-3}]	Concentration in solid phase [mg·g^{-1}]	Adsorbent coefficient [%]
Cu(II)	158.9	15.7	99.4
	317.7	31.3	99.6
	1270.9	125.5	99.9
	3177.3	313.4	99.0
	6354.5	611.7	96.7
	12709.7	1208.2	95.8
Mn(II)	137.4	13.6	99.9
	275.2	27.3	99.9
	1098.1	108.6	99.6
	2746.9	265.6	97.4
	5490.6	519.5	95.5
	10987.8	1035.6	95.2
Pb(II)	518.0	50.4	98.3
	1036.0	100.9	97.9
	4143.3	411.6	99.9
	10360.3	999.8	97.6
	20716.4	1970.3	96.1
	41441.1	3883.2	94.6
Zn(II)	163.6	15.9	98.1
	327.8	31.7	97.4
	1307.4	130.1	99.8
	3268.9	312.4	96.1
	6536.9	605.8	93.9
	13075.4	1204.8	92.9

It was found that the adsorption process of heavy metal ions on the geopolymer is characterized by high efficiency in the concentration ranges studied, as evidenced by the obtained values of the adsorption coefficients for individual ions.

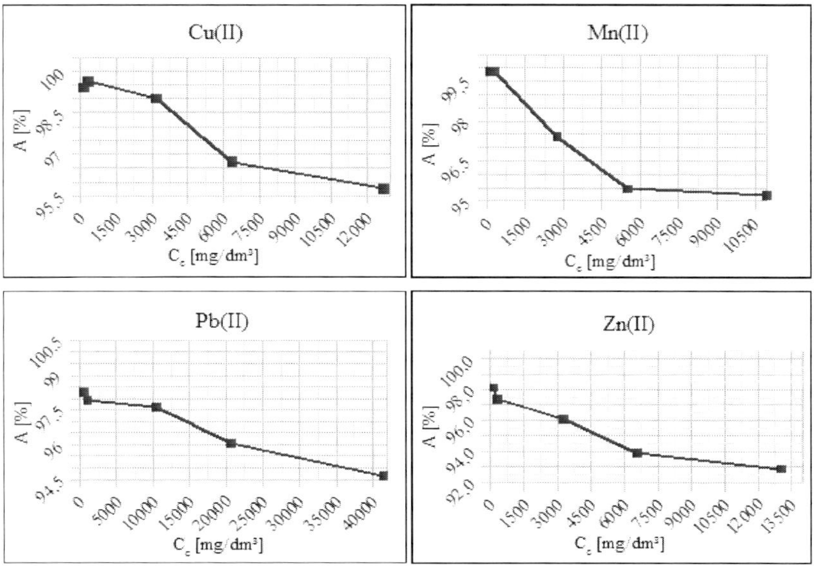

Figure 1. Dependence of the initial concentration of metal ions (Co) on the adsorption coefficient (A) for the adsorption of Cu(II), Mn(II), Pb(II), and Zn(II) by geopolymers (t = 24 h, pH = 4).

Depending on the initial concentration, the value of the adsorption coefficient varies, respectively, for copper(II) in the range 99.9 – 95.8%, for manganese(II) in the range 99.9 – 95.2%, for lead(II) in the range 99.9 –94.6%, and for zinc(II) in the range 99.8 – 92.9%. As the concentration of Cu(II), Mn(II), Pb(II) and Zn(II) ions increases, the value of the adsorption coefficient decreases (Figure 1) and the highest values (99.5 – 99.9%) were found for the following initial concentrations of metal ions studied: 1270.9 – Cu, 1098.1 – Mn, 4143.3 – Pb, and 1307.4 $mg \cdot dm^{-3}$ – Zn. The adsorption efficiency decreases as the initial concentration of the solution used during the experiment increases because at low concentrations more active centers located on the surface of the geopolymers are available. An increase in concentration decreases the number of active sites with which interaction can occur because some of them are already occupied. In addition, increasing the concentration of metal ions in solution increases the competition for active centers, which in turn results in a decrease in the efficiency of the process, as well as a slower establishment of equilibrium [54].

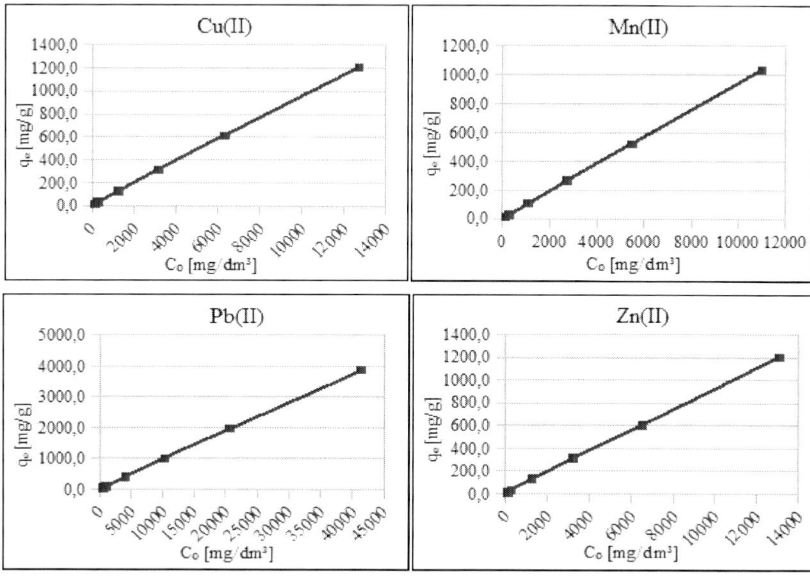

Figure 2. Dependence of the initial concentration of Cu(II), Mn(II), Pb(II), and Zn(II) ions in the solid phase (qe) on the initial concentration of the metal salt solution (Co) for geopolymers (t = 24 h, pH = 4).

Figure 2 shows the dependence of the concentration of ions in the solid phase on the initial concentration of the metal salt solution. It was found that as the initial concentrations of Cu(II), Mn(II), Pb(II), and Zn(II) solutions increased, their solid phase ion concentrations increased. The solid phase ion concentrations for each of the metals studied change in the following series: Mn(II) < Cu(II) < Zn(II) < Pb(II).

The geopolymers studied obtained from biomass combustion ash are characterized by a high removal efficiency of ions from the studied metals Cu(II), Mn(II), Pb(II) and Zn(II). Table 3 presents a comparison of the sorption capacities of selected sorbents. For the metakaolin based geopolymer, the highest sorption capacity was obtained for lead(II) ions in the range of 136.5 to 261.2 mg/g. The lowest sorption capacity was obtained for copper(II) ions, that is, 25.6 – 43.2 mg/g. Kaolinite exhibits the highest sorption capacity for Pb(II) ions, 118.6 mg/g, and the lowest for Cu(II) ions, 0.8 mg/g. For the geopolymer obtained on the basis of fly ash, only sorption capacities for copper(II) and lead(II) ions are given in the literature. For Cu(II) ions, the sorption capacity ranges from 113.4 to 152.3 mg/g, while for Pb(II) ions it ranges from 24.6 to 118.6 mg/g. The experimentally obtained sorption capacities are equal to, respectively: Cu(II) – 133.6, Mn(II) – 104.9, Pb(II) –

348.1, Zn(II) – 132.1 mg/g. Only for copper(II) ions the obtained value coincides with the range reported in the literature. In the case of zinc(II) and lead(II) ions, the values of the sorption capacities given in the study significantly exceed the range in the literature, probably because of the wide range of metal concentrations used in the study.

Table 3. Comparison of sorption capacity values for selected sorbents

Sorbent	Sorption capacity[mg/g]				Literature data
	Cu(II)	Mn(II)	Pb(II)	Zn(II)	
Geopolymer from metakaolin	-	-	-	60.01	[41]
	25.6-43.2	-	136.6-261.2	18.7-35.9	[44]
Geopolymer from fly ash	152.3	-	-	-	[41]
	-	-	24.6-118.6	-	[49]
	113.4	-	-	-	[55]
	133.6	104.9	348.1	132.1	in this study
Glauconitr	2.9	-	9.1	3.1	[56]
	-	-	10.5	1.4	[57]
Kaolinite	1.2	-	2.35		[58]
	4.5	-	-	-	[59]
	1.9	-	-	-	[60]
	0.8	-	-	1.8	[61]
	-	-	11.5	-	[62]
Siderite	-	-	10.3	-	[63]
Montmorylonite	7.2	-	-	-	[64]
Natural zeolite	3.4	-	-	5.4	[65]
Klinoptylolite	1.6		1.6	0.5	[66]
	3.8		6.0	2.7	[67]

Effect of Contact Time on Adsorption Efficiency

The dependence of contact time between the adsorbate and the adsorbent in the range of 5 to 1440 minutes on the adsorption of the studied metal ions with initial concentrations equal to, respectively: 3177 mg·dm^{-3} – Cu(II), 2747 mg·dm^{-3} – Mn(II), 10360 mg·dm^{-3} –Pb(II), and 3269 mg·dm^{-3} – Zn(II). Measurements were made after 1, 5, 15, 30, 60, 180, 300 and 1440 minutes. The results obtained are shown in Table 4 and Figure 3.

It was found that the efficiency of the adsorption process and the equilibrium concentration of metal ions increase with time. The adsorption efficiency of the studied metal ions in geopolymers expressed by the adsorption coefficient after 24 h of contact between the adsorbate and adsorbent changes in the following series:

Pb(II) (88.6%) < Zn(II) (90.6%) < Mn(II) (92.7%) < Cu(II) (95.6%).

The adsorption rate of the studied metal ions on the geopolymer can be described in two stages. In the first stage, a rapid binding of metal ions occurs, which is related to the surface negative charge of the geopolymer, which attracts metal cations. On the other hand, in the second stage, the metal ion binding process is stable because of the saturation of the active centers present on the surface of the geopolymer.

Table 4. Effect of contact time on adsorption of test metal ions by geopolymers from biomass combustion ash (m = 1.00 g, V= 50 cm^3, pH = 4, T = 20°C)

Metal ion	Time [min]	Concentration after time t [$mg \cdot dm^{-3}$]	Concentration in solid phase [$mg \cdot g^{-1}$]	Adsorption coefficient [%]
Cu(II)	5	1567.0	80.7	50.8
	15	778,.	120.2	75.6
	30	422.5	137.9	86.7
	60	322.3	142.9	89.9
	180	173.0	150.4	94.6
	300	160.1	151.1	95.0
	1440	142.2	152.0	95.6
Mn(II)	5	1633.4	57.6	41.4
	15	1027.5	87.9	63.1
	30	707.0	103.9	74.6
	60	600.9	109.2	78.5
	180	387.2	119.9	86.1
	300	240.0	127.2	91.4
	1440	201.7	129.1	92.7
Pb(II)	5	6967.5	169.7	32.8
	15	5300.4	253.1	48.9
	30	4167.4	309.7	59.8
	60	2750.2	380.5	73.5
	180	2000.3	418.1	80.7
	300	1433.3	446.4	86.2
	1440	1183.1	458.9	88.6
Zn(II)	5	1383.4	94.7	57.8
	15	867.8	120.6	73.6
	30	700.7	128.9	78.6
	60	5505.0	136.4	83.2
	180	333.2	147.2	89.8
	300	317.0	148.1	90.3
	1440	308.0	148.5	90.6

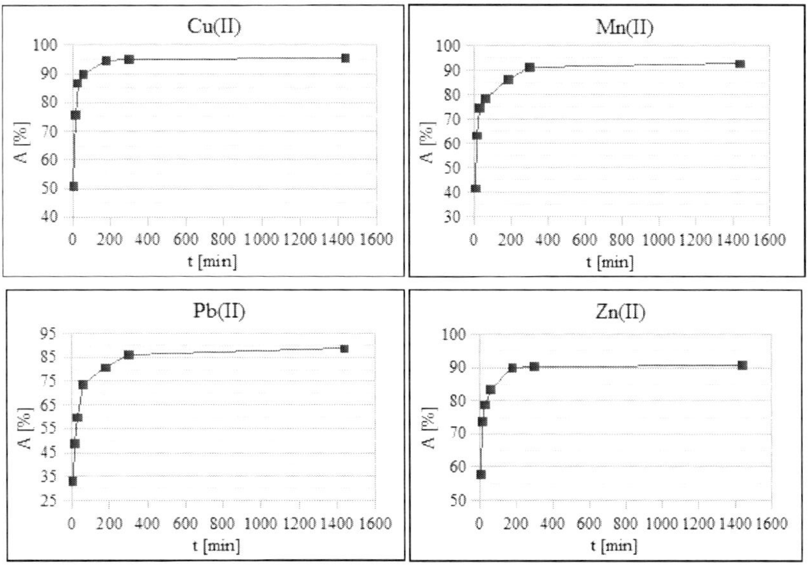

Figure 3. Dependence of contact time on the adsorption coefficient for ions of the metals studied with initial concentrations: 3177 - Cu(II), 2747 - Mn(II), 10360 - Pb(II), and 3269 mg·dm^{-3} - Zn(II).

Desorption of Test Metal Ions

The stability of binding between Cu(II), Mn(II), Pb(II), and Zn(II) metal cations and the geopolymer obtained from biomass ash was determined by the degree of desorption. The results of the study of the effect of initial concentration of metals on their geopolymer desorption process are summarized in Table 5.

For the metal ions studied, the desorption coefficient changes in the following ranges for copper(II) 14.2 - 28.6%, manganese(II) 12.5 - 21.8%, lead(II) 12.8 - 29.4%, and zinc(II) 13.7 - 32.0%. As the concentration of Cu(II), Mn(II), Pb(II) and Zn(II) ions increases, the value of the desorption coefficient increases (Table 5) and the highest values (21.8 - 32.0%) were found for the following initial concentrations of the studied metal ions: 12709.7 – Cu(II), 10987.8 – Mn(II), 41441.1 – Pb(II), and 13075.4 mg·dm^{-3} – Zn(II).

Table 5. Effect of the initial concentration of the studied metal ions on their desorption of geopolymers (t = 24 h, T = 20°C)

Metal ion	Initial concentration [mg·dm^{-3}]	Concentration after dsorption [mg·dm^{-3}]	Desorption coefficient [%]
Cu(II)	158.9	22.6	14.2
	317.7	60.7	19.1
	1270.9	310.1	24.4
	3177.3	835.6	26.3
	6354.5	1766.6	27.8
	12709.7	3635.0	28.6
Mn(II)	137.4	17.2	12.5
	275.2	40.7	14.8
	1098,1	202.1	18.4
	2747,0	450.5	16.4
	5490,6	1065.2	19.4
	10987.8	2395.3	21.8
Pb(II)	518.1	66.3	12.8
	1036.1	161.6	15.6
	4143.3	696.1	16.8
	10360.3	2838.7	27.4
	20716.4	5282.7	25.5
	41441.1	12183.7	29.4
Zn(II)	163.3	22.4	13.7
	327.8	64.2	19.6
	1307.4	377.8	28.9
	3268.8	882.6	27.0
	6537.0	1935,.0	29.6
	13075.4	4184.0	32.0

Description of the Equilibrium and Kinetics of the Adsorption of Cu(II), Mn(II), Pb(II), and Zn(II) In Geopolymers Obtained from Biomass Combustion Ashes

Description of the Quilibrium of Adsorption of Cu(II), Mn(II), Pb(II), and Zn(II) in Geopolymers Obtained from Biomass Combustion Ashes

The literature mainly uses Langmuir and Freundlich models [41, 42, 49, 54] to describe the relationship between the amount of a substance adsorbed on a geopolymer and its concentration in the equilibrium model solution. Modeling the equilibrium of adsorption for the metal ions studied is shown in Figures 4-7 sequentially.

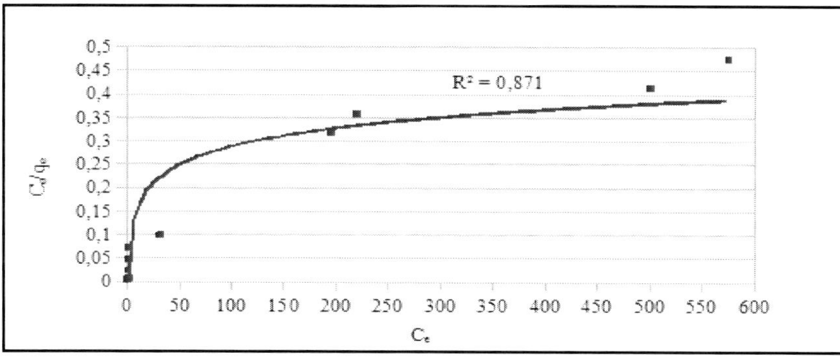

(a)

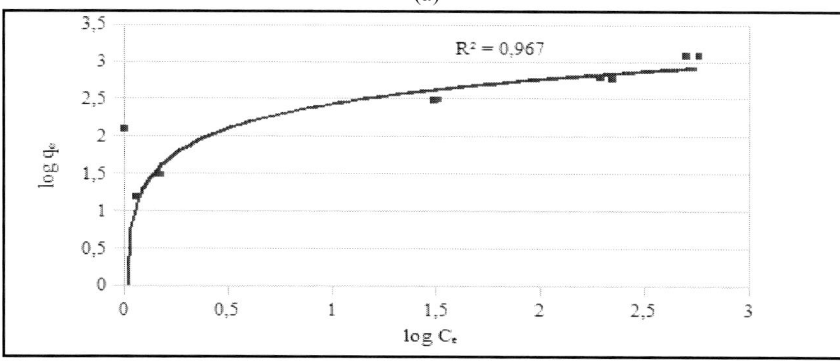

(b)

Figure 4. Modeling of the adsorption equilibrium for Cu(II) ions by the (a) Langmuir and (b) Freundlich isotherms.

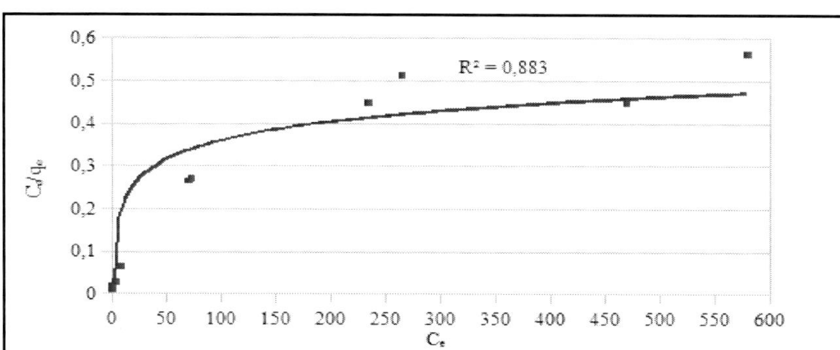

(a)

Figure 5. (Continued)

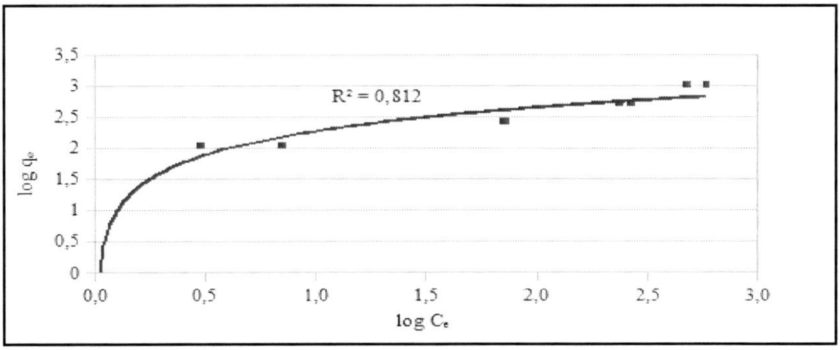

(b)

Figure 5. Modeling of the adsorption equilibrium for Mn(II) ions by the (a) Langmuir and (b) Freundlich isotherms.

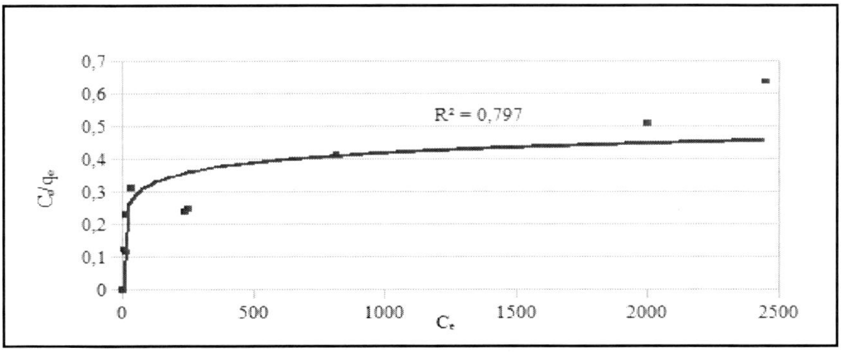

(a)

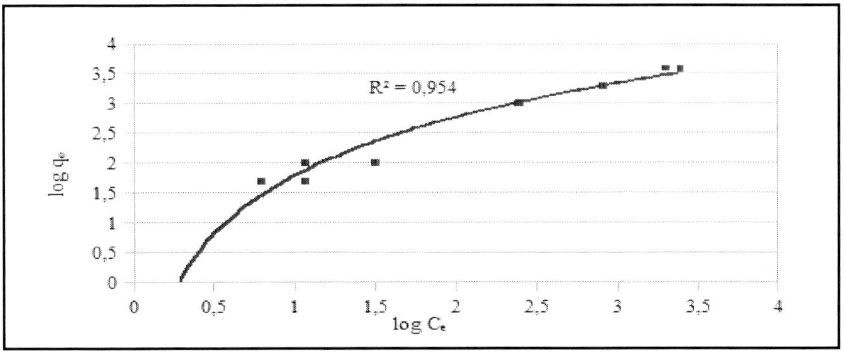

(b)

Figure 6. Modeling of the adsorption equilibrium for Pb(II) ions by the (a) Langmuir and (b) Freundlich isotherms.

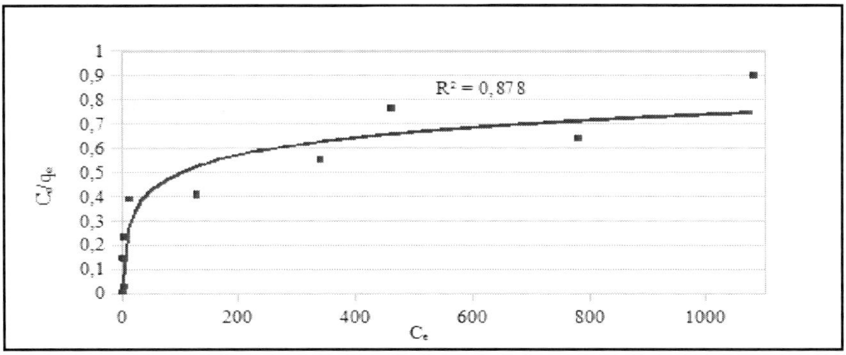

(a)

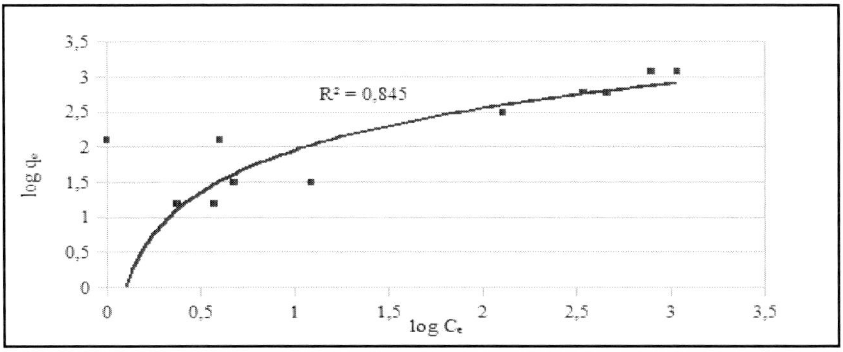

(b)

Figure 7. Modeling of the adsorption equilibrium for Zn(II) ions by the (a) Langmuir and (b) Freundlich isotherms.

Table 6. Summary of the parameters of Langmuir and Freundlich isotherm models for the studied metal ions

Metal ion	Langmuir		Freundlich		
	K_L [dm^3/mg]	R^2	K_F [mg/g]	1/n	R^2
Cu (II)	0.058	0.871	0.482	2.430	0.967
Mn (II)	0.064	0.883	0.552	2.267	0.812
Zn (II)	0.106	0.878	0.869	1.949	0.845
Pb (II)	0.043	0.797	1.415	1.785	0.954

Based on the sorption isotherms obtained from the studied metal ions, which are shown in Figures 4-7, and the knowledge of the values of the linear correlation coefficients R^2 (summarized in Table 6), it is possible to fit the appropriate models to describe the equilibrium of adsorption. In the case of Cu(II) and Pb(II) ions, the sorption process is better described by the

Freundlich model, as evidenced by the high value of R^2. Both isotherm models can be used to describe the adsorption equilibrium of Mn(II) and Zn(II) ions as they show similar values of linear correlation coefficients.

Description of the Kinetics of the Adsorption of Cu(II), Mn(II), Pb(II), and Zn(II) in Geopolymers Obtained from Biomass Combustion Ashes

The results obtained for the effect of contact time on the adsorption of Cu(II), Mn(II), Pb(II), and Zn(II) on geopolymers obtained from biomass combustion ashes were used to determine the adsorption kinetics. The study of the adsorption kinetics of Cu(II), Mn(II), Pb(II), and Zn(II) ions on geopolymers obtained from biomass combustion ashes was performed using pseudo-first order and pseudo-second order kinetics equations.

In the literature, pseudo-first order and pseudo-second order kinetics equations are most commonly used to study the adsorption of heavy metal ions on combustion by-products [68, 69]. The adsorption kinetics is dependent on the interactions between the adsorbate and the adsorbent and the state in which the system is in.

The kinetic model of the pseudo-first order equation, can be defined as Equation (3) [69]:

$$\ln(q_e\text{-}q_t) = \ln q_e\text{-}k_1 t \qquad (3)$$

where: q_e - equilibrium adsorption capacity [$mg{\cdot}g^{-1}$], q_t - number of adsorbed ions [$mg{\cdot}g^{-1}$], k_1 - rate constant [1/min], t - time [min].

From this equation a relationship can be determined that describes the reaction rates from the concentration of individual reactants. It shows a satisfactory fit for equilibrium concentrations with lower values, and its use is limited to processes in which diffusion through the boundary layer occurs prior to adsorption and where adsorption exhibits a physical character.

The kinetic model of the pseudo-second order equation was defined in terms of Equation (4) [69]:

$$\frac{t}{q_t} = \frac{1}{k_2 \cdot q_e^2} + \frac{1}{q_e} \qquad (4)$$

where: q_e - equilibrium adsorption capacity [$mg{\cdot}g^{-1}$], q_t - number of adsorbed ions [$mg{\cdot}g^{-1}$], k_2 - rate constant [$(g{\cdot}min){\cdot}mg^{-1}$], t - time [min].

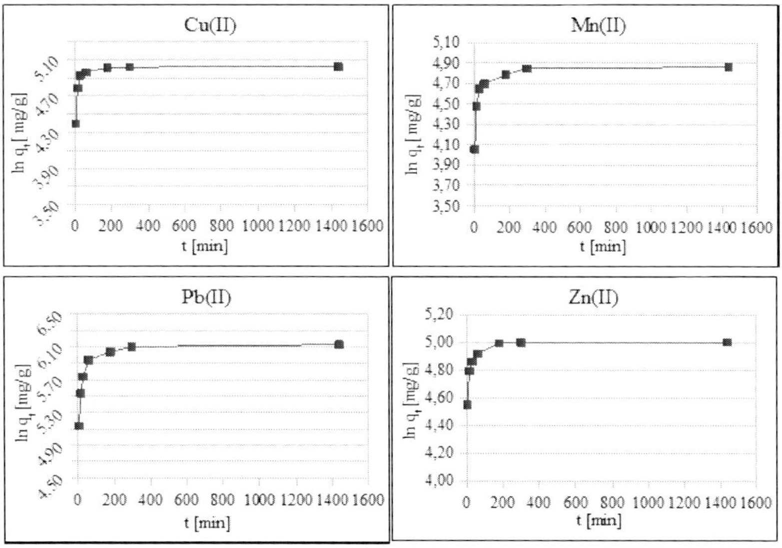

Figure 8. Fitting of pseudo-first order kinetics equation to experimental data for adsorption of Cu(II), Mn(II), Pb(II), and Zn(II) ions on geopolymers (q_t - ion concentration in the solid phase after time t).

The linearized integral form for the pseudo-second order kinetic model can be represented as Equation (5) [69]:

$$\frac{t}{q_t} = f(t) \tag{5}$$

where: q_t - number of adsorbed ions [$mg{\cdot}g^{-1}$], t - time [min].

Figures 8 and 9 show the fit of the pseudo-first order and pseudo-second order kinetics equation to the experimental data for the adsorption of the studied metal ions. The experimental data are better described by the pseudo-second order equation because the determined linear regression equations have higher coefficients of determination (of the order of 0.9999), compared to those for the pseudo-first order equation (0.1612-0.2839). High coefficients of determination for the pseudo-second order equation indicate chemisorption. The pseudo-second order model assumes that the rate-limiting step in the process is chemisorption, indicating that the adsorption of metal ions on the geopolymer occurs through chemisorption [49]. Literature reports [41, 49, 54] confirm that the adsorption process of metal ions on geopolymers obtained from waste materials and aluminosilicate minerals occurs according to the pseudo-second order model.

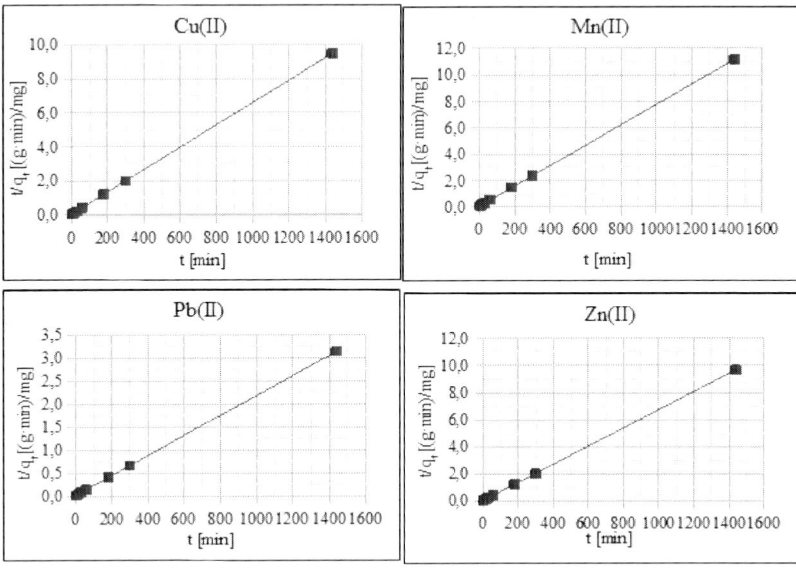

Figure 9. Fitting of pseudo-second order kinetics equation to experimental data for adsorption of Cu(II), Mn(II), Pb(II), and Zn(II) ions on geopolymers (q_t - ion concentration in solid phase after time t).

Table 7 summarizes the calculated kinetic parameters for the pseudo-first order and pseudo-second order equations for the geopolymers used.

Table 7. Kinetic parameters for pseudo-first and pseudo-second order kinetics equation

Metal ion	Pseudo-first order equation		Pseudo-second order equation	
	R^2	k_1 [min^{-1}]	R^2	k_2 [(g·min)·mg^{-1}]
Cu(II)	0.1612	$1.69 \cdot 10^{-2}$	0.9999	$2.58 \cdot 10^{-5}$
Mn(II)	0.2466	$1.60 \cdot 10^{-2}$	0.9999	$3.29 \cdot 10^{-5}$
Pb(II)	0.2839	$2.02 \cdot 10^{-2}$	0.9999	$9.91 \cdot 10^{-6}$
Zn(II)	0.2308	$1.68 \cdot 10^{-2}$	0.9999	$2.61 \cdot 10^{-5}$

The rate constants determined from the pseudo-second order equation are in the range of 9.91-10^{-6} to 3.29-10^{-5} [(g·min)·mg^{-1}]. Taking into account the adsorption rate, the studied metal ions can be ranked in the following order:

Pb (II) < Cu (II) < Zn (II) < Mn (II)

Based on the pseudo-second order equation, the equilibrium concentrations for the studied ions were calculated, which are, respectively:

Cu(II) - 128.19, Mn(II) - 101.09, Pb(II) - 336.71, Zn(II) - 126.91 mg·g^{-1}. The calculated values are close to the experimental values, equal to, respectively: Cu(II) - 133.6, Mn(II) - 104.9, Pb(II) - 348.1, Zn(II) - 132.1 mg·g^{1}, which confirms their reliability.

Study of Geopolymers Obtained from Biomass Combustion Ashes After Adsorption of the Studied Metal Ions

Structural Parameters

Figure 10 shows the low temperature nitrogen adsorption isotherms for geopolymers obtained from biomass combustion ash after adsorption of the studied ions: a) Cu(II), b) Mn(II), c) Pb(II), and d) Zn(II). Their shape corresponds to the type IV isotherm according to the IUPAC classification, which is characteristic of mesoporous adsorbents. Additionally, Figure 10 shows adsorption hysteresis (type H3), which is related to the different course of isotherms of adsorption (when the pressure increases) and desorption (when the pressure decreases) isotherms. The type of hysteresis is characteristic of pores shaped like open double-sided slits or cylinders [70].

The BET adsorption isotherm for the relative pressure range of 0.05 to 0.25 mm Hg is shown in Figure 11, from which the specific surface area was determined for the geopolymers obtained from the biomass ash after the adsorption of the metal ions studied. The small specific surface area and the pore volume indicate that the geopolymer matrix is dense. This means that the above-mentioned geopolymers are characterized by low permeability, and consequently, high durability [71].

Table 8 shows the porosimetric results, that is, the specific surface area, volume, and pore size of the geopolymers studied. Comparing all geopolymers after adsorption of the Cu(II), Mn(II), Pb(II), and Zn(II) ions tested (denoted Cu/GEO, Mn/GEO, Pb/GEO, and Zn/GEO) in terms of the obtained BET surface areas, it was found that for Pb/GEO the smallest BET surface area was 9.070 m^2/g, while for Zn/GEO it is 68.203 m^2/g. The total micropore area determined by the t method for the different geopolymers after adsorption of the studied metal ions is: 3.792 m^2/g for Cu/GEO, 4.598 m^2/g for Mn/GEO, 0.599 m^2/g for Pb/GEO and 2.479 m^2/g for Zn/GEO, which is respectively: 21.38%, 16.37%, 6.60%, 3.63% of the specific surface area.

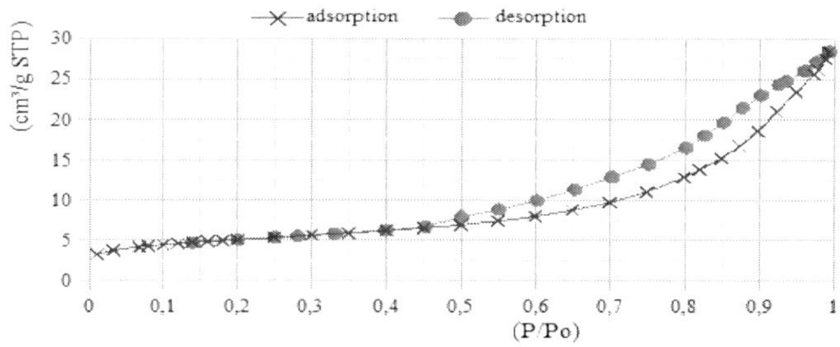

(a)

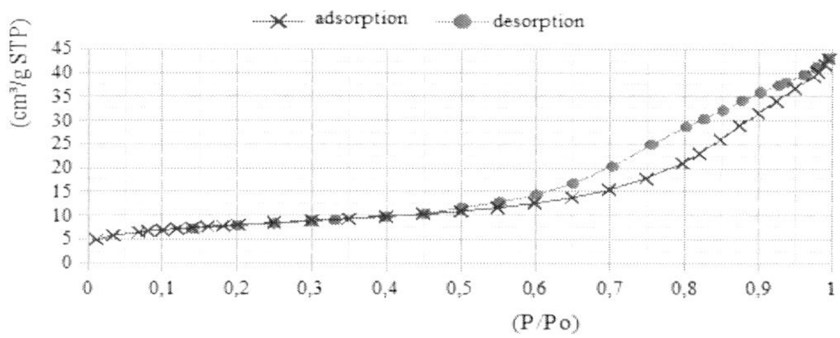

(b)

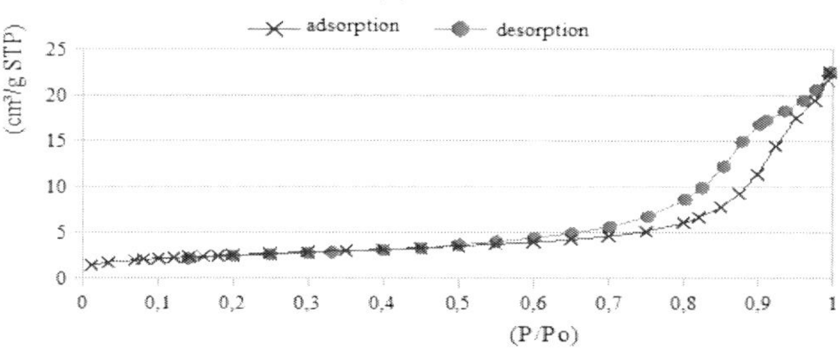

(c)

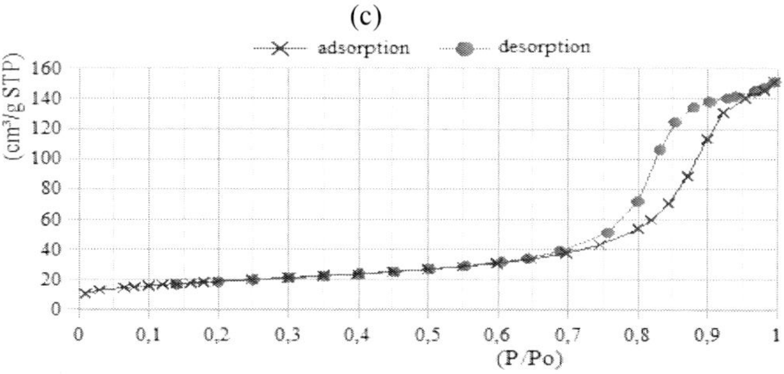

(d)

Figure 10. Nitrogen adsorption and desorption isotherms for geopolymers obtained from biomass combustion ash after adsorption of: a) Cu(II), b) Mn(II), c) Pb(II), and d) Zn(II) ions determined by the BET method.

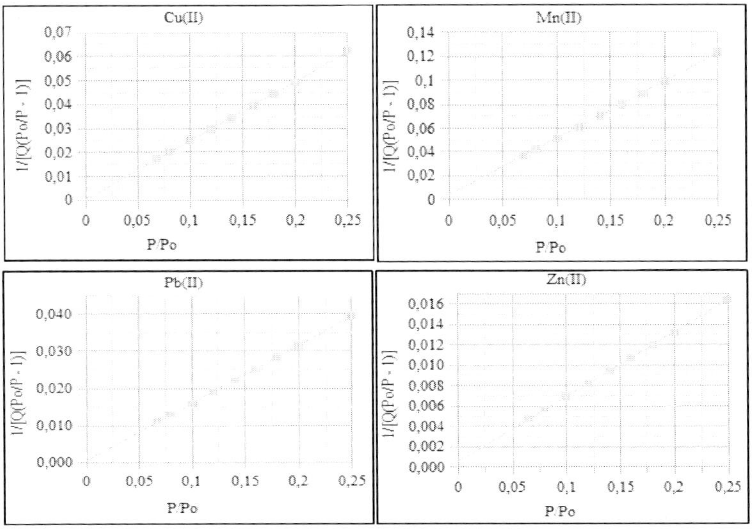

Figure 11. Linear form of the BET adsorption isotherm for geopolymers obtained from biomass combustion ash after the adsorption of Cu(II), Mn(II), Pb(II), and Zn(II) ions.

Table 8. Porous structure parameters of geopolymers after adsorption of Cu (II)/GEO, Mn (II)/GEO, Pb (II)/GEO and Zn (II)/GEO determined by low temperature nitrogen adsorption and desorption

	Cu/GEO	Mn/GEO	Pb/GEO	Zn/GEO
Surface area [m^2/g]				
Single point surface area at $P/P_o = 0.1$	17.47	27.49	8.79	65.91
BET Surface Area	17.74	28.08	9.07	68.20
Langmuir Surface Area	24.81	39.42	12.81	95.80
t-Plot External Surface Area	3.79	4.60	0.60	2.48
Pore volume [cm^3/g]				
Single point desorption total pore volume of pores less than 805.413 Å diameter at P/P_o = 0.1	0.04	0.06	0.03	0.23
BJH Adsorption cumulative volume of pores between 17.000 Å and 3000.000 Å diameter	0.04	0.06	0.03	0.23
BJH Desorption cumulative volume of pores between 17.000 Å and 3000.000 Å diameter	0.05	0.07	0.04	0.23
Pore size [Å]				
Desorption average pore width (4V/A by BET)	94.63	90.45	140.74	133.83
BJH Adsorption average pore diameter (4V/A)	106.35	100.22	160.03	124.69
BJH Desorption average pore diameter (4V/A)	86.52	84.93	124.66	105.29

Surface Morphology and Chemical Composition Analysis

Figure 12 shows the microstructure of the geopolymer obtained from biomass combustion ash after adsorption of Cu(II) ions. A large surface heterogeneity can be observed, which is related to the various grain structures of the biomass ash and the presence of amorphous silicates, which have a porous surface. The results of the EDS analysis for the geopolymer obtained from biomass combustion ash after adsorption of Cu(II) ions are shown in Figure 13. The lowest copper content adsorbed on the surface of the geopolymer obtained from biomass combustion ash was found in area 1, which has the highest silicate ion content. The highest copper content is found in amorphous porous forms, where the silicate ion content is the lowest. Oxygen, silicon, and copper were found in all areas, with an additional presence of calcium in area 3. The dominant share in all areas is represented by oxygen (35.3 - 47.3%). The copper content in individual areas varies from 6.2 to 55.8%.

Figure 12. SEM image of geopolymer obtained from biomass combustion ash after the adsorption of Cu(II) ions.

Figure 14 shows the microstructure of the geopolymer obtained from biomass combustion ash after adsorption of Mn(II) ions. As in the case of Cu(II) ions, a large surface heterogeneity can be observed, which is related to the different grain structures of the biomass ash and the presence of amorphous silicates, which have a porous surface. The presence of oxygen, magnesium, aluminum, silicon, calcium, manganese, and iron was found in the studied areas. In all areas, the presence of oxygen, silicon, potassium, calcium, and manganese was found, in area 2 also the presence of aluminum. The dominant share in all areas shows oxygen (22.7 - 46.4%). The manganese content in individual areas varies from 14.7 to 16.4%, with the lowest content of this metal found on the surface of the spherical forms of approximately 4% (Figure 15).

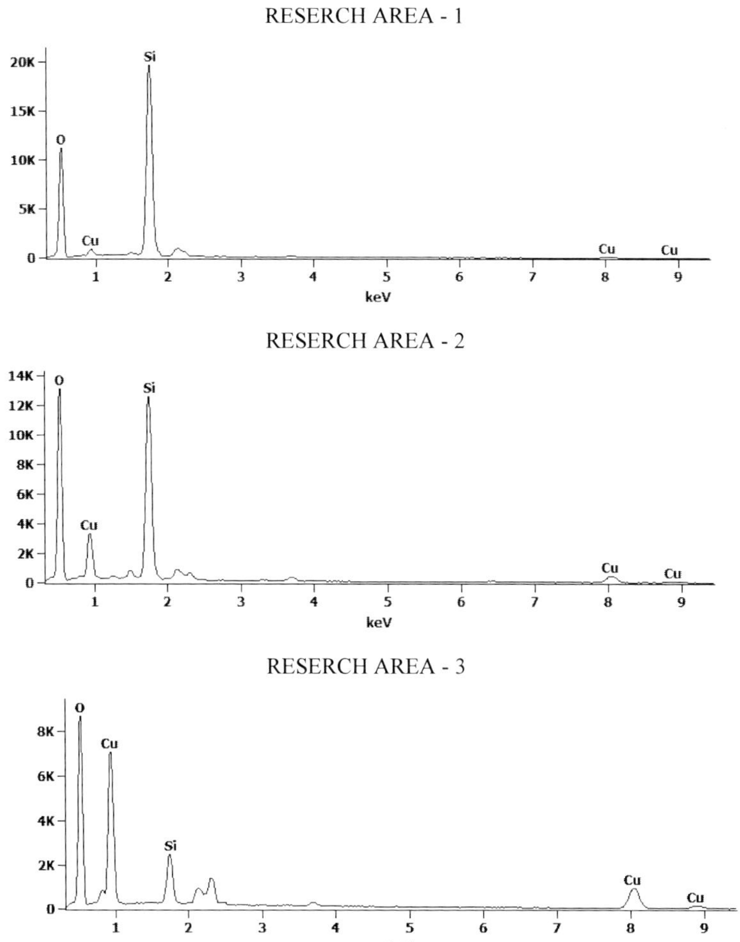

Figure 13. EDS analysis (research areas 1-3) for the geopolymer obtained from biomass combustion ash after adsorption of Cu(II) ions.

Figure 14. SEM image of geopolymer obtained from biomass combustion ash after the adsorption of Mn(II) ions.

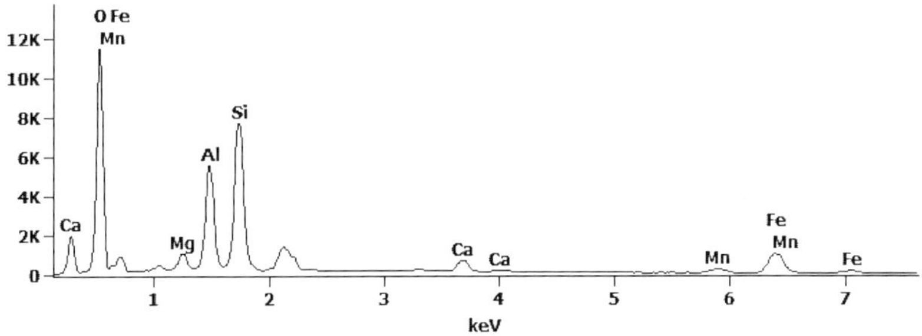

Figure 15. EDS analysis (research areas 1) for the geopolymer obtained from biomass combustion ash after adsorption of Mn(II) ions.

Figure 16 shows the microstructure of the geopolymer obtained from biomass combustion ash after adsorption of Pb(II) ions. A large surface heterogeneity can be observed, which is related to the different grain structures of the biomass ash and the presence of amorphous silicates, which have a porous surface. The results of the EDS analysis for the geopolymer obtained from biomass combustion ash after adsorption of Pb(II) ions are shown in Figure 17. The dominant contribution in all areas shows oxygen (26.0 - 47.2%). The lead content in individual areas varies from 7.1 - 50.0%. The highest content of Pb(II) ions is found in amorphous porous forms, where the content of silicon and aluminum ions is the lowest.

Figure 16. SEM image of geopolymer obtained from biomass combustion ash after the adsorption of Pb(II) ions.

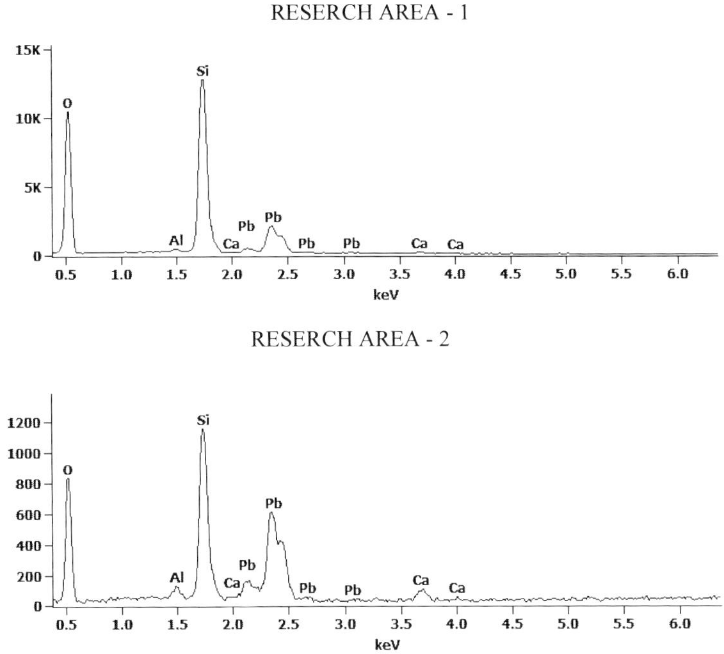

Figure 17. (Continued)

RESERCH AREA - 3

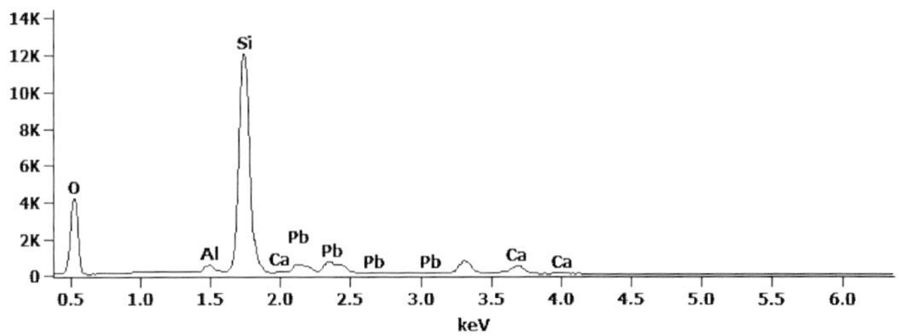

RESERCH AREA - 4

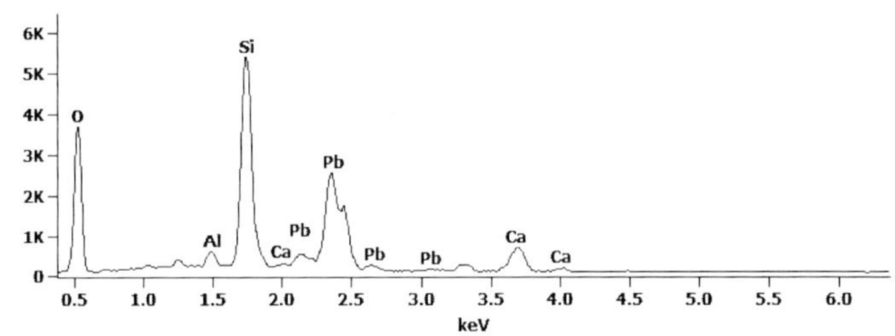

RESERCH AREA - 5

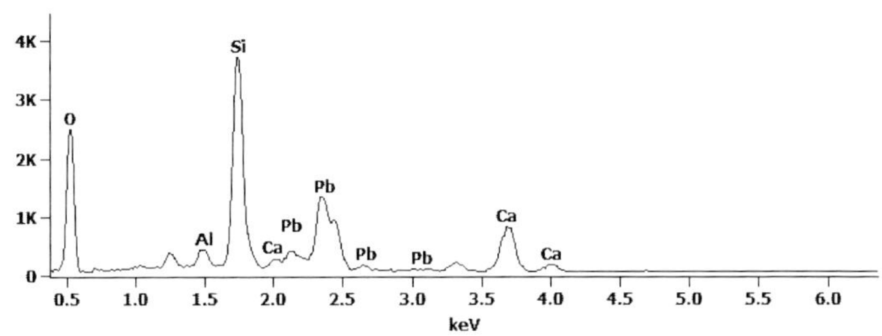

RESERCH AREA - 6

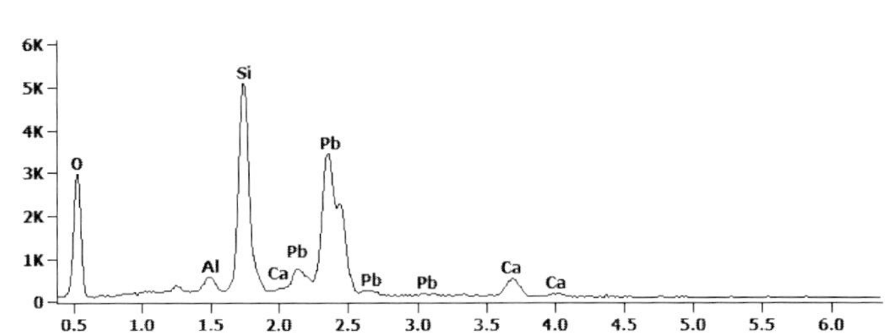

Figure 17. EDS analysis (research areas 1-6) for the geopolymer obtained from biomass combustion ash after adsorption of Pb(II) ions.

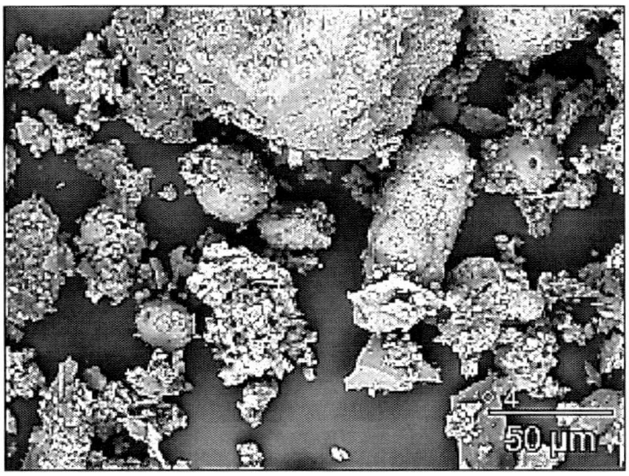

Figure18. SEM image of geopolymer obtained from biomass combustion ash after the adsorption of Zn(II) ions.

Figure 18 shows the microstructure of the geopolymer obtained from biomass combustion ash after adsorption of Zn(II) ions. A large surface heterogeneity can be observed, which is related to the different grain structures of the biomass ash and the presence of amorphous silicates, which have a porous surface. The results of the EDS analysis for the geopolymer obtained from biomass combustion ash after adsorption of Pb(II) ions are shown in Figure 19. Oxygen, aluminum, silicon, potassium, iron and zinc were found in all areas, in areas 3 - 4 additionally the presence of calcium, in area 1 the presence of sodium. Oxygen (19.8 - 46.2%) and silicon (15.1 - 31.0%) show the dominant share in all areas. The zinc content in individual areas varies from 3.2 to 52.5%. The lowest content of Zn(II) ions adsorbed on the surface of the geopolymer obtained from biomass combustion ash was found in spherical forms, which have the highest content of silicate ions and aluminum ions. The highest content of Zn(II) ions is found in amorphous porous forms, where the content of silicate and aluminum ions is the lowest.

Spectroscopic Analysis

The sorption capacity of geopolymers is due to the presence of characteristic functional groups, which are capable of binding to a selected group of pollutants from aqueous systems. In aqueous media, functional groups such as hydroxyl and carbonyl dissociate, and the adsorbent surface has a negative charge and can adsorb metal cations [72, 73]. Heavy metal cations can be immobilized on geopolymers by two mechanisms, i.e., ion exchange and chemisorption [74].

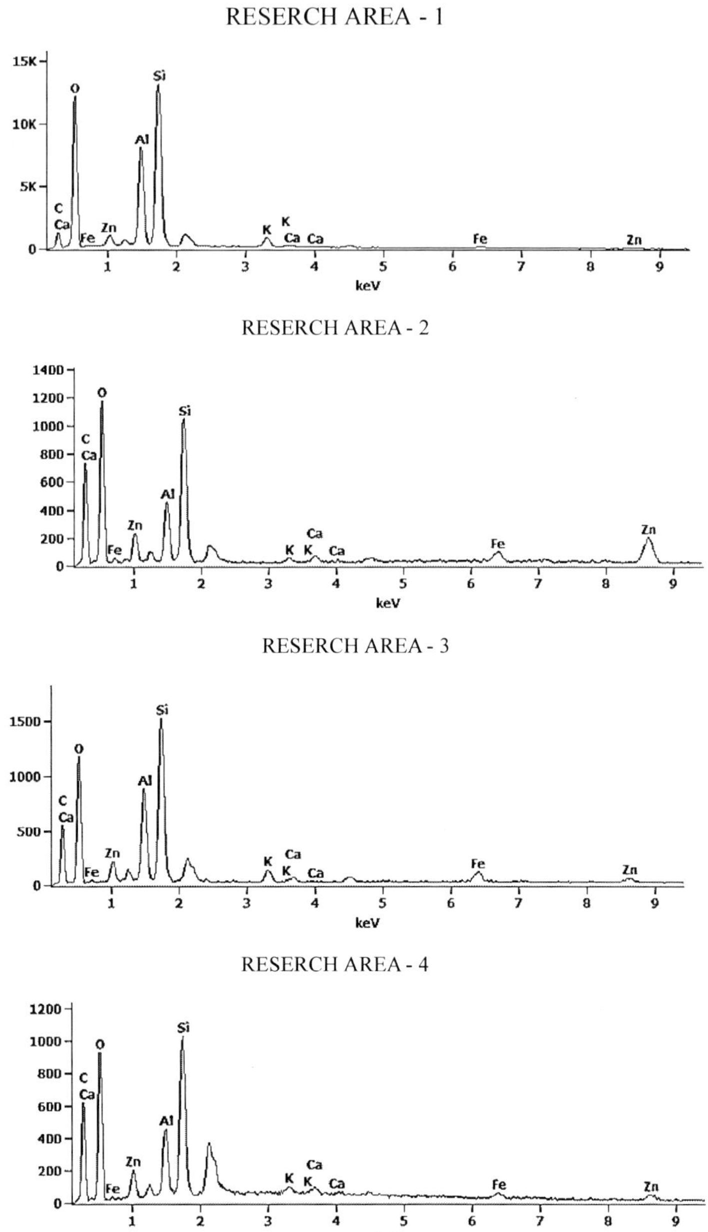

Figure 19. EDS analysis (research areas 1-4) for the geopolymer obtained from biomass combustion ash after adsorption of Zn(II) ions.

The ion exchange was estimated from the value of the desorption coefficient (Table 5), while the chemisorption was calculated as the difference between the total sorption value, i.e., 100%, and the ion exchange. For each of the test metal ions, the sorption process was found to occur mainly according to the chemisorption mechanism. The average proportion of chemisorption is higher compared to the ion exchange for each of the studied metal ions and is as follows: for Cu(II) - 98.15%, Mn(II) - 98.50%, Pb(II) - 98.32% and Zn(II) - 97.58%, respectively.

Spectroscopic results were also used to describe the sorption mechanism. The spectra of geopolymer samples were compared after adsorption and desorption of the studied metal ions. The visible differences are mainly related to the change in intensity and shifts of the bands associated with O-H, C-O, and N-H stretching vibrations, which indicates their involvement in the adsorption process of the studied metal ions. Moreover, the type of functional groups involved in the adsorption process of the individual metal ions was determined. The spectra obtained after adsorption are shown in Figures 20 - 23, and the characteristics of the individual bands are summarized in Table 9. On the other hand, the obtained spectra after desorption are shown in Figures 24 - 27, and the characteristics of the individual bands are summarized in Table 10.

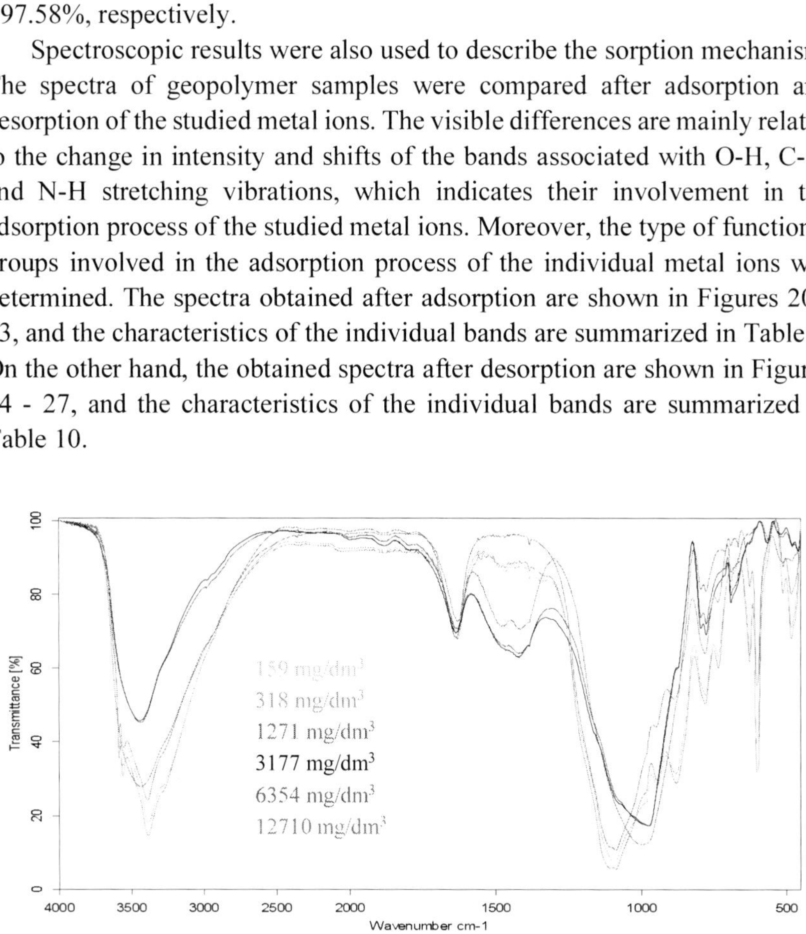

Figure 20. FTIR spectra of geopolymers obtained after adsorption of Cu(II) ions from aqueous solutions with concentrations ranging from 159 to 12710 mg·dm^{-3}.

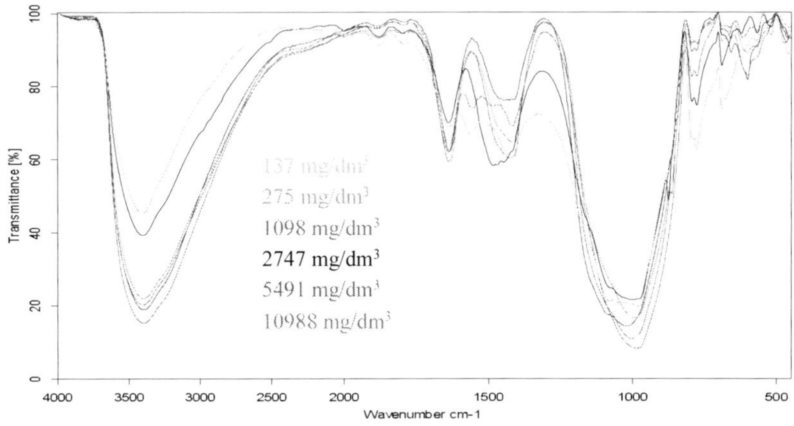

Figure 21. FTIR spectra of geopolymers obtained after adsorption of Mn(II) ions from aqueous solutions with concentrations ranging from 137 to 10988 mg·dm^{-3}.

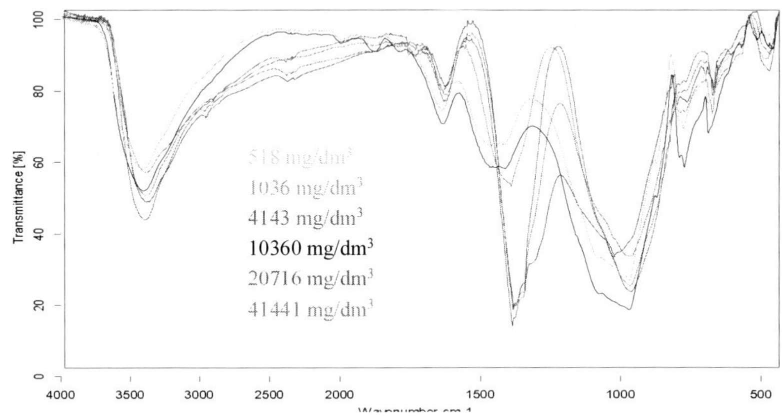

Figure 22. FTIR spectra of geopolymers obtained after adsorption of Pb(II) ions from aqueous solutions with concentrations ranging from 518 to 41441 mg·dm^{-3}.

Table 9. Characteristics of individual bands after adsorption of the ions studied from aqueous solutions

Band position [cm^{-1}]	Types of vibration
3450	Stretching vibrations of the OH and NH groups
1650 – 1630	Tensile vibrations of the -NH_2 group
1450 – 1400	Tensile vibrations of the CO group
1000 – 500	Asymmetric vibrations of Si-O(Si) and Si-O(Al) binding and internal vibrations characteristic of aluminosilicates

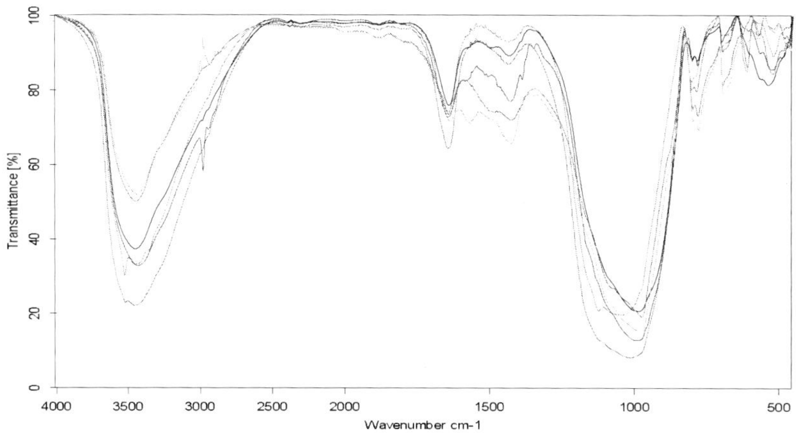

Figure 23. FTIR spectra of geopolymers obtained after adsorption of Zn(II) ions from aqueous solutions with concentrations ranging from 164 to 13075 mg·dm^{-3}.

Table 10. Characteristics of individual bands after desorption of studied ions from aqueous solutions

Band position [cm^{-1}]	Types of vibration
3490 – 3300	Stretching vibrations of theOH and NH groups
1670 – 1620	Tensile vibrations of the -NH_2 group
1420 – 1300	Tensile vibrations of the CO group
1100 – 500	Asymmetric vibrations of Si-O(Si) and Si-O(Al) binding and internal vibrations characteristic of aluminosilicates

From the comparison of the presented FTIR spectra of geopolymers obtained after adsorption of Cu(II), Mn(II), Pb(II), and Zn(II) ions using different initial concentrations of metal ions, it was observed that the recorded changes are related to four areas, viz:

- 3600 - 3200 cm^{-1} - associated with OH and NH stretching vibrations,
- 1650 - 1560 cm^{-1} - derived from stretching vibrations - NH_2,
- approximately 1400 cm^{-1} - associated with CO stretching vibrations,
- 1100 - 500 cm^{-1} - coming from Si-O(Si) and Si-O(Al) vibrations and internal vibrations characteristic for aluminosilicates.

Visible changes in the position and intensity of individual bands indicate that during the adsorption of Cu(II), Mn(II), Pb(II), and Zn(II) ions on geopolymer based on biomass ash, bonds between functional groups, i.e., O-H, N-H, and C-O, with cations of the above-mentioned metals are formed.

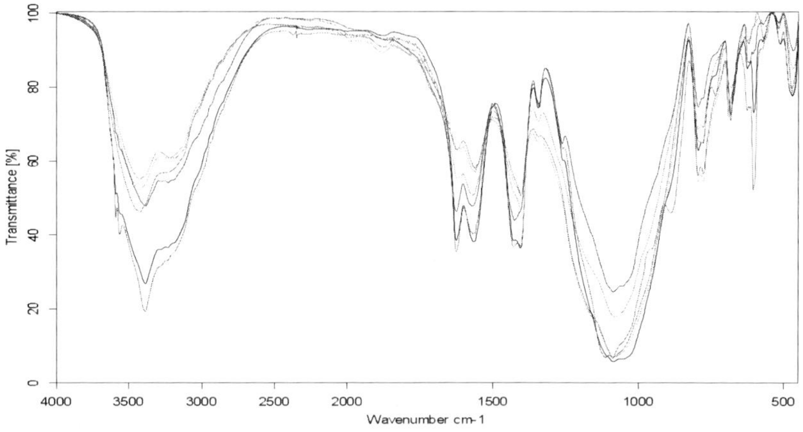

Figure 24. FTIR spectra of geopolymers obtained after desorption of Cu(II) ions from aqueous solutions with concentrations ranging from 159 to 12710 mg·dm^{-3}.

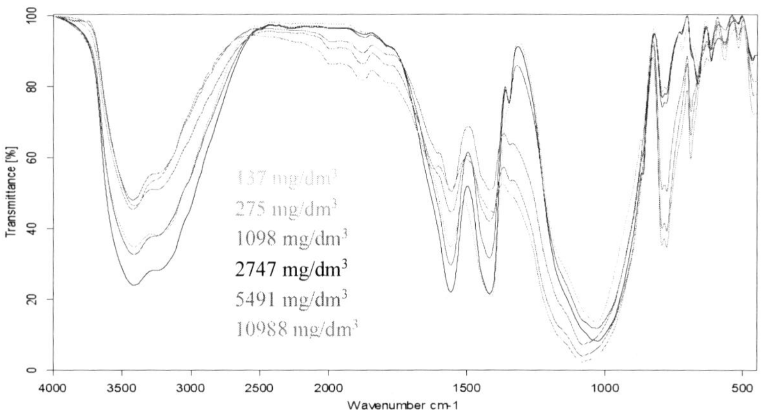

Figure 25. FTIR spectra of geopolymers obtained after desorption of Mn(II) ions from aqueous solutions with concentrations ranging from 137 to 10988 mg·dm^{-3}.

From the comparison of the presented FTIR spectra of geopolymers obtained after desorption of Cu(II), Mn(II), Pb(II) and Zn(II) ions using different initial concentrations of metal ions, it was observed that the recorded changes are related to four areas, viz:

- 3490 - 3300 cm^{-1} - related to OH and NH stretching vibrations,
- 1670 - 1560 cm^{-1} - derived from stretching vibrations - NH_2,
- 1450 - 1300 cm^{-1} - associated with CO stretching vibrations,
- 1100 - 500 cm^{-1} - from Si-O(Si) and Si-O(Al) vibrations and internal vibrations characteristic for aluminosilicates.

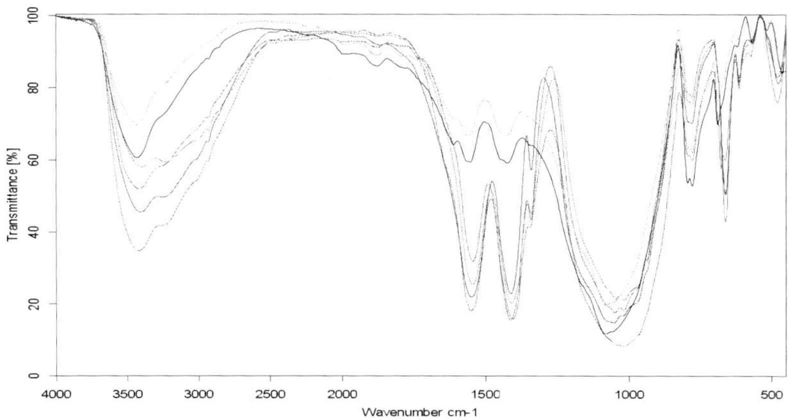

Figure 26. FTIR spectra of geopolymers obtained after desorption of Pb(II) ions from aqueous solutions with concentrations ranging from 518 to 41441 mg·dm^{-3}.

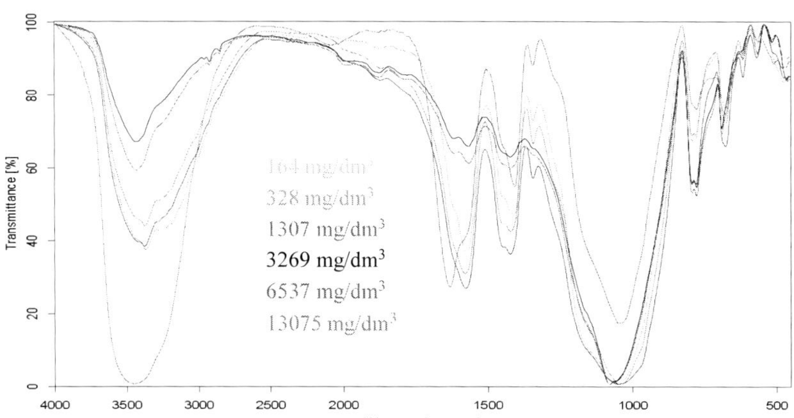

Figure 27. FTIR spectra of geopolymers obtained after desorption of Zn(II) ions from aqueous solutions with concentrations ranging from 164 to 13075 mg·dm^{-3}.

The apparent changes in the position and intensity of the individual bands indicate that chemisorption of the studied ions Cu(II), Mn(II), Pb(II) and Zn(II) occurs on the biomass ash-based geopolymer during the desorption process.

Conclusion

Based on the adsorption studies performed of Cu(II), Mn(II), Pb(II) and Zn(II) ions on geopolymers obtained from biomass combustion ash, the following conclusions can be drawn:

1. The adsorbent used shows high adsorption efficiency of Cu(II), Mn(II), Pb(II) and Zn(II) ions in the studied model systems, respectively:

   98.4 - Cu(II), 97.9 - Mn (II), 97.4 - Pb(II), and 96.37% - Zn(II).

2. The adsorption process is well described by the Freundlich isotherm, which describes adsorption on heterogeneous surfaces. The equilibrium parameters indicate that the geopolymers used have good adsorption capacity, and the process occurs by chemisorption.
3. The adsorption process efficiency and equilibrium concentration of metal ions were found to increase with time. The adsorption efficiency of the studied metal ions in geopolymers expressed by the adsorption coefficient after 24 h of contact between the adsorbate and the adsorbent can be represented as follows:

   Pb(II) (88.6%) < Zn(II) (90.6%) < Mn(II) (92.7%) < Cu(II) (95.6%).

4. The experimentally obtained sorption capacities for the studied metal ions are equal to, respectively:

   Cu(II) - 133.6, Mn(II) - 104.9, Pb(II) - 348.1, Zn(II) - 132.1 mg/g^1, respectively.

5. The adsorption of the investigated Cu(II), Mn(II), Pb(II), and Zn(II) ions on geopolymers obtained from biomass combustion ashes follows the pseudo-second order model. The determined rate constants indicate that the adsorption rate occurs in the following series:

   Pb(II) < Cu(II) < Zn(II) < Mn(II).

6. The analysis of FTIR spectra of geopolymers after adsorption of the studied metal ions indicates the involvement of functional groups i.e., O-H, N-H, and C-O in the adsorption process, which confirms that the adsorption process of Cu(II), Mn(II), Pb(II), and Zn(II) ions occurs mainly by chemical sorption mechanism.

References

[1] Vassilev S. V., Baxter D., Andersen L. K., Vassileva C. G. (2010). An overview of the chemical composition of biomass. *Fuel* 89:913-933.

[2] Ryu C., Bin Yang Y., Kho A., Yates N. E., Nsharifi V., Swithenbank J. (2006). Effect of fuel properties on biomass combustion: Part I. Experiments – fuel type, equivalence ratio and particle size. *Fuel*, 85:1039-1046.

[3] Vassilev S. V., Baxter D., Andersen L. K., Vassileva C. G. (2013). An overview of the behaviour of biomass during combustion: Part I. Phase-mineral transformations of organic and inorganic matter. *Fuel* 112:391-449.

[4] Reis J. S., Araujo R. O., Lima V. M. R., Queiroz L. S., da Costa C. E. F., Pardauil J. J. R., Chaar J. S., Rocha Filho G. N., de Souza L. K. C. (2019). Combustion properties of potential Amazon biomass waste for use as fuel. *Journal of Thermal Analysis and Calorimetry* 138:3535-3539.

[5] Girón R. P., Ruiz B., Fuente E., Gil R. R., Suárez-Ruiz I. (2013). Properties of fly ash from forest biomass combustion. *Fuel* 114:71-77.

[6] Uliasz-Bocheńczyk A., Pawluk A., Pyzalski M. (2016). Characteristics of ashes from biomass combustion in fluidized bed boilers. *Mineral Resources Management* 32:149-162.

[7] Wójcik M., Stachowicz F., Masłoń A. (2017). Possibility to use of fly ash to improve sewage sludge dewatering. *Journal of Civil Engineering, Environment and Architecture* 1:377-393.

[8] James A., Thring R., Helle S., Ghuman H. (2012). Ash management review - applications of biomass bottom ash. *Energies* 5:3856-3873.

[9] Kosior-Kazberuk M. (2011). New mineral additives for concrete. *Construction and Environmental Engineering* 1: 47-55.

[10] Brzozowski P. (2011). Possibilities of using fly ash from combustion in fluidized bed boilers for concrete laid under water. *Construction and Environmental Engineering* 1:1-11.
[11] Ahmaruzzaman M. (2010). A review on the utilization of fly ash, *Progress in Energy and Combustion Science* 36:327-363.
[12] Fernández-Pereira C., de la Casa J., Gómez-Barea A., Arroyo F., Leiva C., Luna Y. (2011). Application of biomass gasification fly ash for brick manufacturing. *Fuel* 90:220-232.
[13] Franus W., Wdowin M. (2011). The use of class F fly ash for the production of zeolite material on a semi-technical scale. *Energy Politic* 14:79-91.
[14] Zhang X. Y., Chen L., Komarneni S., Zhou C. H., Tong D. S., Yang H. M., Yu W. H., Wang H. (2016). Fly ash-based geopolymer: clean production, properties and applications. *Journal of Cleaner Production* 125:253-267.
[15] Mikuła J. (ed.) (2014). A pro-ecological solution in the field of production. Modern environmentally friendly composite materials. Publishing House of the Cracow University of Technology, Kraków (in polish).
[16] García R., Pizarro C., Álvarez A., Lavín A. G., Bueno J. L. (2015). Study of biomass combustion wastes. *Fuel* 148:152-159.
[17] Embong R., Kusbiantoro A., Shafiq N., Nuruddin M. F. (2016). Strength and microstructural properties of fly ash based geopolymer concrete containing high-calcium and water-absorptive aggregate. *Journal of Clenear Production* 112:816-822.
[18] Palomo A., Alonso S., Fernandez-Jimenez A., Sobrados L., Sanz J. (2004). Alkaline activation of fly ash. A NMR study of the reaction products. *Journal of the American Ceramic Society* 87:1141-1145.
[19] Bajare D., Bumanis G., Shakhmenko G., Justs J. (2011). Obtaining composition of geopolymers (alkali activated binders) from local industrial wastes, Paper presented at the 3rd International Conference Civil Engineering, *Proceedings Building Materials* 50-56.
[20] Rajamma R., Labrincha J. A., Ferreira V. M. (2012). Alkali activation of biomass fly ash–metakaolin blends. *Fuel* 98:265-271.
[21] Candamano S., De Luca P., Frontera P., Crea F. (2017). Production of geopolymeric mortars containing forest biomass ash as partial replacemeng of metakaolin. *Environments* 4:1-13.
[22] Kaur K., Singh J., Kaur M. (2018). Compressive strength of rice husk ash based geopolymer: The effect of alkaline activator. *Construction and Building Materials* 169:188-192.
[23] Sturm P., Gluth G. G., Brouwers H., Kühne H. (2016). Synthesizing one-part geopolymers from rice husk ash. *Construction and Building Materials* 124:961-966.
[24] Detphan S., Chindaprasirt P. (2009). Preparation of fly ash and rice husk ash geopolymer. *International Journal of Minerals, Metallurgy and Materials* 16:720-726.
[25] Samadhi T. W., Wulandari W., Prasetyo M. I., Fernando M. R., Purbasari A. (2017). Synthesis of geopolymer from biomass-coal ash blends. *AIP Conference Proceedings* 1887, 020031.

[26] He J., Jie Y., Zhang J., Yu Y., Zhang G. (2013). Synthesis and characterization of red mud and rice husk ash-based geopolymer composites. *Cement and Concrete Composites* 37:108-118.
[27] Lemougna P. N., Wang K., Tang Q., Melo U. C., Cui X. (2016). Recent developments on inorganic polymers synthesis and applications. *Ceramics International* 42:15142-15159.
[28] Davidovits J. (1993). Geopolymer cement to minimize carbon-dioxide greenhouse-warming, in cement-based materials: Present, Future and Environmental Aspects. *Ceramic Transactions* 37:165-182.
[29] Brylicki W., Stryczek S., Gonet A. (2006). Geopolymer grouts for injection work. *Oil Gas Drilling* 23:121-129.
[30] Zhang Z., Wang K., Mo B., Li X., Cui X. (2015). Preparationandcharacterization of a reflective and heat insulative coating based on geopolymers. *Energy Build* 87:220-225.
[31] Wiyono D., Antoni A., Hardjito D. (2015). Improving the durability of pozzolan concrete using alkaline solution and geopolymer coating. *Procedia Engineering* 125:747-753.
[32] Zarina Y., Al Bakri M. A. M, Kamarudin H., Khairul Nizar I., Rafiza A. R. (2013). Reviews on the Geopolymer Materials for Coating Application. *Advanced Materials Research* 626:958-962.
[33] Temuujin, J., Minjigmaa A., Rickard W., Lee M., Williams I., van Riessen A. (2009). Preparation of metakaolin based geopolymer coatings on metal substrates as thermal barriers. *Applied Clay Science* 46:265-270.
[34] Sazama P, Bortnovsky O., Dědečeka J., Tvarůžková Z., Sobalíka Z. (2011). Geopolymer based catalysts - new group of catalytic materials. *Catalysis Today* 164:92-99.
[35] Mikuła J., Łach M., Mierzwiński D. (2017). Ways of managing ashes and slags from waste incineration plants. *Ecological Engineering* 18:37-46.
[36] Zhang J., Provis J. L., Feng D., van Deventer J. (2008). Geopolymers for stabilization of Cr^{6+}, Cd^{2+} and Pb^{2+}. *Journal of Hazardous Materials* 157:587-598.
[37] El-Eswed B., Yousef R., Alshaaer M., Hamdnehd I., Al-Gharabli S., Khalili F. (2015). Stabilization/solidification of heavy metals in kaolin/zeolite based geopolymers. *International Journal of Mineral Processing* 137:34-42.
[38] El-Eswed B., Aldagag O., Khalili F. (2017). Efficiency and mechanism of stabilization/solidification of Pb(II), Cd(II),Cu(II), Th(IV) and U(VI) in metakaolin based geopolymers. *Applied Clay Science* 140:148-156.
[39] Siyal A. A., Shamsuddin M. R., Khan M. I., Rabat N. E., Zulfiqar M., Man Z., Siame J., Azizli K. A. (2018). A review on geopolymers as emerging materials for the adsorption of heavy metals and dyes. *Journal of Environmental Management* 224:327-339.
[40] Al-Zboon K., Al-Harahshed M. S., Hani F. B. (2011). Fly ash-based geopolymer for Pb removal from aqueous solution. *Journal of Hazardous Materials* 188:414-421.
[41] Kara I., Yilmazer D., Akar S. T. (2017). Metakaolin based geopolymer as an effective adsorbent for adsorption of zinc(II) and nickel(II) ions from aqueous solutions. *Appied Clay Science* 139:54-63.

[42] Cheng T., Lee M., Ko M., Ueng T., Yang S. (2012). The heavy metal adsorption characteristics on metakaolin-based geopolymer. *Applied Clay Science* 56:90-96.

[43] López F., Sugita S., Tagaya M., Kobayashi T. (2014). Metakaolin-based geopolymers for targeted adsorbents to heavy metal ion separation. *Journal of Materials Science and Chemical Engineering* 2:16-27.

[44] Andrejkovičová S., Sudagar A., Rocha J., Patinha C., Hajjaji W., Ferreirada Silva E., Velosa A., Rocha F. (2016). The effect of natural zeolite on microstructure, mechanical and heavy metals adsorption properties of metakaolin based geopolymers. *Applied Clay Science* 126:141-152.

[45] Singhal A., Gangwar B. P., Gayathry J. M. (2017). CTAB modified large surface area nanoporous nanoporous. *Applied Clay Science* 150:106-114.

[46] Naghsh M., Shams K. (2017). Synthesis of a kaolin-based geopolymer using a novel fusion method and its application in effective water softening. *Applied Clay Science* 146:238-245.

[47] Sarkar C., Basu J. K., Samanta A. N. (2017). Removal of Ni^{2+} ion from waste water by geopolymeric adsorbent derived from LD Slag. *Journal of Water Process Engineering* 17:237-244.

[48] Javadian H., Ghorbani F., Tayebi H., Asl S. (2015). Study of the adsorption of Cd(II) from aqueous solution using zeolite-based geopolymer, synthesized from coal fly ash; kinetic, isotherm and thermodynamic studies. *Arabian Journal of Chemistry* 8:837-849.

[49] Liu Y., Yan C., Zhang Z., Wang H., Zhou S., Zhou W. (2016). A comparative study on fly ash, geopolymer and faujasite block for Pb removal from aqueous solution. *Fuel* 185:181-189.

[50] Lee N. K., Khalid H. R., Lee H. K. (2017). Adsorption characteristics of cesium onto mesoporous geopolymers containing nano-crystalline zeolites. *Microporous and Mesoporous Materials* 242:238-244.

[51] Sitarz-Palczak E., Kalembkiewicz J., Galas D. (2019). Comparative study on the characteristics of coal fly ash and biomass ash geopolymers. *Archives of Environmental Protection* 45:126-135.

[52] Kalembkiewicz J., Galas D., Sitarz-Palczak E. (2018). A Novel Class of Industrial Ash-Based Geopolymer: Synthesis, Properties and Fractionation of Selected Heavy Metals. Municipal and Industrial Waste: Sources, *Management Practices and Future Challenges,* NOVA SCIENCE Publishers, New York, 157-180.

[53] Kutcho B., Kim A. (2006). Fly ash characerization by SEM/EDS. *Fuel* 85:2537-2544.

[54] Al-Harahsheh M. S., Zboon K. A., Al-Makhadmeh L., Hararah M., Mahasneh M. (2015). *Fly ash based geopolymer for heavy metal removal: A case study on copper.*

[55] Duan P., Yan C., Zhou W., Ren D. (2016). Development of fly ash and iron ore tailing based porous geopolymer for removal of Cu(II) from wastewate. *Ceramics International* 42:13507-13518..

[56] Franus M., Bandura L. (2014). Sorption of heavy metal ions from aqueous solution by glauconite. *Fresenius Environmental Bulletin* 23:825-839.

[57] Lu E., Smith W., Vengris T., Binkiene R. (1996). Sorption of heavy metals by Lithuanian glauconite. *Water Research* 30:2883-2892.

[58] Jin M., Lu X., Jiang X., Chen Z. (2010). Adsorption of Pb(II), Cd(II), Ni(II) and Cu(II) onto natural kaolinite clay. *Desalination* 52:33-39.

[59] Ulmanu M., Maranon E., Fernandez Y., Castrillon L., Anger I., Dumitriu D. (2002). Removal of copper and cadmium ions from diluted aqueous solutions by low cost and waste material adsorbents. *Water Air and Soil Pollution* 42:357–373.

[60] Arias M., Barral M. T., Mejuto J. C. (2002). Enhancement of copper and cadmium adsorption on kaolin by the presence of humic acids. *Chemosphere* 48:1081–1088.

[61] Chantawong V., Harvey N. W., Bashkin V. N. (2003). Comparison of heavy metal adsorptions by Thai kaolin and ball clay. *Water Air Soil Pollution* 148:111-125.

[62] Gupta S. S., Bhattacharyya K. G. (2008). Immobilization of Pb(II), Cd(II) and Ni(II) ions on kaolinite and montmorillonite surfaces from aqueous medium. *Journal of Environmental Management* 87:46-58.

[63] Erdem M., Ozverdi A. (2005). Lead adsorption from aqueous solution onto siderite. *Separation and Purification Technology* 42:259–264.

[64] Ijagbemi C. O., Baek M. H., Kim D. S. (2009). Montmorillonite surface properties and sorption characteristics for heavy metal removal from aqueous solutions. *Journal of Hazardous Materials* 166:538–546.

[65] Motsi T., Rowson N. A., Simmon M. J. H. (2009). Adsorption of heavy metals from acid mine drainage by natural zeolite. *International Journal of Mineral Processing* 92:42-48.

[66] Zamzow M. J., Eichbaum B. R., Sandgren K. R., Shanks D. E. (1990). Removal of heavy metals and other cations from wastewater using zeolites. *Separation Science and Technology* 25:1555–1569.

[67] Elliot H. A., Huang C. P. (1981). Adsorption characteristics of some Cu(II) complexes on aluminosilicates. *Water Research* 15:849–855.

[68] Chojnacka K., Michalak I. (2009). Using wood and bone ash to remove metal ions from solutions. *Global NEST Journal* 11:205-217.

[69] Xu L., Cui H., Zheng X., Liang J., Xing X., Yao L., Chen Z., Zhou J. (2017). Assorption of Cu^{2+} to biomass ash its modified product" *Water Science and Technology* 1:115-125.

[70] Donohue M. D., Aranovich G. L. (1998). Adsorption hysteresis in porous solids". *Journal of Colloid and Interface Science* 205:121-130.

[71] Alehyen S., Zerzouri M., Lalouani E., Achouri M., Taibi M. (2017). Porosity and fire resistance of fly ash based geopolymer. *Journal of Materials and Environmental Sciences* 8:3676-3689.

[72] Bailey S. E., Olin T. J., Bricka R. M., Adrian D. D. (1999). A review of potentially low-cost sorbents for heavy metals. *Water Research* 33:2469-2479.

[73] Li H., Shen X. Z., Sisk B., Orndorff W., Li D., Pa W. P., Riley J. T. (1997). Studies of fly ash using thermal analysis techniques. *Journal of Thermal Analysis and* Calorimetry 49:943-951.

[74] Mozgawa W., Król M., Bajda T. (2009). Application of IR spectra in the studies of heavy metal cations immobilization on natural sorbents. *Journal of Molecular Structure* 924-926: 427-433.

Chapter 3

Geo-Polymer Concretes: Structure, Properties and Materials

Berivan Yilmazer Polat, PhD
Munzur University, Tunceli, Türkiye

Abstract

In recent years, different types of special concretes have been produced and used according to the usage principles of the buildings. One of these concretes is alkali active concretes produced without the need for cement production which is known as geopolymer concrete. Their high early strength, less drying-shrinkage, high fire, abrasion & acid resistance, less CO_2 emissions, least setting time, and high freeze-thaw & corrosion resistance are reasons this type of concrete has emerged as an alternative to portland cement concrete. This concrete is formed by the reaction of material rich in alumina and silica with the activator. Since geopolymer concrete is more environmentally friendly than cementitious concrete, it has been the subject of many studies. Geopolymer concretes have some disadvantages despite all these advantages. One of them is brittleness. It is known that adding fibers to geopolymer concretes positively affects brittleness and ductility, compressive strength, and flexural strength. Also, mineral additives strengthen the durability of concrete. This work presents a summary of the literature on geopolymer concrete's foundations and disadvantages compared to portland cement concrete, mix proportions that give the best strength, management of fiber additives, performance analysis against burst rate, and resistance to attack.

Keywords: geopolymer concrete, geo-concrete, alkali-active concrete, green concrete

In: Geopolymers
Editor: 979-8-88697-197-2
ISBN: Joseph E. Glenn

Introduction

Concrete is one of the most demanding construction materials. Due to the demand for concrete in the construction sector, the production of ordinary portland cement (OPC) is carried out very intensively [1]. OPC production causes environmental pollution because approximately one ton of OP production releases one ton of CO_2 into the atmosphere [2–6]. Developments in science and technology have led to innovations and advances in the construction sector. One of these innovations is geopolymer concrete.

Davidovits is the first person to do work on geopolymer cement. Davidovits proposed that an alkaline liquid could be used to react with silica (Si) and aluminum (Al) in by-product materials such as fly ash and rice husk ash to produce binders, or in the material of geological origin. Since this chemical reaction is a polymerization process, the term 'geopolymer' has been introduced to the literature to represent these binders. Geopolymer is an inorganic polymeric material prepared by activating aluminosilicate materials such as fly ash or metakaolin with activators that become reaction products by polycondensation and segmented dissolution in alkaline conditions. Geopolymer concretes, which can reduce carbon dioxide emissions by up to 80% and whose carbon footprint is approximately 9% less than ordinary portland cement concrete (OPCC), do not need high-temperature furnaces during production and do not require the use of excess fuel. Thus, it has a positive effect on global warming [7–11]. Due to the air pollution and harmful effects caused by cement production, these concretes, which can be considered relatively environmentally friendly, have created an alternative that can be used instead of cement-based binders. Geopolymer concretes provide advantages such as high early strength, low shrinkage, freeze-thaw resistance, sulfate and corrosion resistance, and long-term retention of toxic and hazardous wastes [12]. Therefore, recently it is especially preferred in the production of materials such as high-performance composites and ceramics [13–16].

Various studies have been conducted to identify the differences between geopolymer concrete and OPCC.

- In a study that compared the permeability of OPCC and geopolymer concrete, Park et al. determined the open cavity volumes and water absorption capacities of these two different concretes. Accompanied by this data, it was revealed that geopolymer concrete had a 37.62%

reduction in open cavity compared to standard concrete and absorbed 38.38% less water [17]. In their study, Joseph and Mathew compared the modulus of elasticity and Poisson ratio of OPCC and geopolymer concrete. According to the OPCC, it was found that geopolymer concrete increased by 14.4% in the modulus of elasticity and 19.2% in the Poisson ratio [18].

Given all these features, the impact experiment and the explosion experiment were considered to research geopolymer concrete as an option for portland cement concrete. In this study, depending on the content of the subject, the literature on geopolymer concretes, impact, and explosion effect on geopolymer concrete will be examined.

Structure of the Geopolymer

Geopolymers were first discovered and used by Prof. J. Davidovits [14], [19]. After its discovery, it was found that geopolymers were used in many ancient structures [20, 21].

Geopolymer chemistry was laid with the discovery of polysialates obtained as a result of siloxane or silicate molecules (AlO4) formed from chain bonds of silica molecules sharing oxygen atom electrons with aluminate molecules [22].

In general, geopolymer belongs to the semi-amorphous inorganic alumina-silicate polymer class, which is synthesized mainly from silica and aluminum materials, such as mineral additives obtained from waste materials [23]. It has no shape compared to other crystalline zeolitic materials [13], [24, 25]. It can be synthesized at 80°C by holding water substances in the geopolymer. But zeolite crystallizes at 150-250 °C with leaving gaps in its structure [22]. In addition, the synthetic structure of geopolymer components is similar to zeolites [26, 27].

The mechanism of geopolymers in concrete occurs when aluminate and silicates in the structure of baked clay form covalent-bonded compounds using oxygen atom electrons in common [28]. Oxidized compounds of Al and Si elements, especially in high amounts in kaolin clay, form monomers that provide binding and volumetric toughness [29]. The most important factor determining the molecular structure is the total Si: Al molar ratio of geological soil source and activating alkaline silicates [29, 30].

Aluminosilicate materials commonly used in the production of geopolymer concrete are fly ash and blast-furnace slag [31]. They produce a polymeric reaction by binding inorganic alumina-silicate polymers synthesized from silicium and aluminum material [25].

Kaolinite and metacaolinite are relatively inexpensive alumina silicates used in most of the earlier work on geopolymerization [23, 32–35].

The most common alkaline solution used in geopolymerization is a compound solution of NaOH/KOH and Na_2SiO_3 or K_2SiO_3 [36, 37]. Alkaline solutions are of great importance in terms of concrete strength. In his study, Rashad observed that the compressive strength of geopolymer concrete increases to a certain point with the concentration of NaOH solution [37, 38].

In geopolymerization, the weight ratio of alumina-silicate powder to the alkaline solution is very high, usually between 3.0 and 5.5 [33, 35].

The process of polymerization consists of dissolution and rearrangement, condensation, and polymerization. This structural model of the geopolymer is presented by Davidovits [27, 39].

The dissolution and rearrangement of aluminosilicate create various oligomers [40]. Oligomers form by binding large polymers [41]. When oligomers bind, the OH groups at their ends meet and release water by sharing an oxygen atom [41].

Geopolymer Concrete Production

Puzolonic Materials

Geopolymer concrete needs different binders compared to OPCC. These binders are alkaline liquids and pozzolanic materials. Pozzolanic materials are rich in silicium (Si) and aluminum (Al) and consist of materials such as metakaolin, fly ash, and slag [42, 43]. Geopolymer concrete is formed by the polymerization of aluminosilicates of pozzolanic materials (Figure 1). In addition, existing agricultural and industrial wastes such as rice husk ash; ash residue can be used to replace OPC and imported additives. This use of cheaper natural pozzolans or waste materials can provide significant savings and reduce the environmental costs caused by mixing cement and greenhouse gases released during the production of portland cement [44, 45]. These aluminosilicates have a significant effect on the physical and mechanical properties of geopolymer concrete [13, 46].

Figure 1. Mineral Additives In Geopolymer Concrete [47].

Metakaolin

Metakaolin has high pozzolanic properties and is produced using high temperatures (500-900°C) to calcinate kaolin, and contains high levels of SiO_2 and Al_2O_3 [48, 49]. Metakaolin, which is conducive to high-strength geopolymer production, is an industrial mineral material with homogeneous properties. It requires a lower temperature than the OPC for its production, so 80-90% less CO_2 emissions occur during its production [28, 50]. For this reason, it is a widely available material [9]. This has brought many studies of metakaolin-based geopolymers into the literature. It is seen that the use of metakaolin in concrete or mortars increases workability and allows for a smoother surface [51, 52] In addition, it has been observed that the use of metakaolin increases the resistance of freezing dissolution in concrete, reduces shrinkage, and increases bending, tensile & compressive strength [49, 53–55].

It is obtained by sintering kaolin clay at temperatures of 600, 700, 800, 900, and 1000ºC.

Alaa et al. concluded that the optimum temperature for heating the kaolin substance to obtain metakaolin could be between 600 and 850°C for 1-12 hours. They argued that the differences found may be related to the chemical composition of metakaolin of kaolin calcination at 650°C. Most researchers have reported that the optimal heating time is 3 hours, and the optimal calcination conditions occur at a temperature of 700°C for 2 hours [56, 57].

As with other mineral additives, metakaolin can act as a binder in cement-free concretes. Binders are a big factor in strength. In other words, materials with high strength and durability properties can be obtained with components containing Al-Si compound in optimum proportions of Metakaolinite clay, which is pulverized by burning F class fly ash or 750ºC with these binders. Non-baked kaolin clay gives lower compressive strength [28, 36].

Some disadvantages of metakaolin have also been reported by researchers. Duxson, Omar, et al. (2018) observed that the properties of metakaolin-based geopolymer concrete are better than other hydrates because it is more permanent, but requires higher water demand despite its advantages [13, 46].

Fly Ash

Fly ash is artificial pozzolans that are not binding on their own, can fly under the influence of temperature, and are in a fine-grained structure used as a mineral additive in concrete formed by the burning of coal [58, 59]. Fly ash becomes 75-80% of the total ash generated by burning coal. Hence, fly ash is a commonly found material [60].

Fly ash is waste produced by thermal reactors during energy production, such as blast-furnace slag and other aluminosilicate materials derived from metakaolin and similar geological sources [61–63]. These wastes are hydrated with slaked lime and hardened in water and increase the strength of concrete against possible effects [58]. It has been observed that geopolymer concretes with fly ash addition have high strength even in an acid ambient, and call at a lower deformation than OPCC [28].

Palomo, Grutzeck, and Blanco found that different fly ash samples that were cured for 24 hours at 85°C had compressive strength between 35 and 40 MPa. But they found that the compressive strength reached about 90 MPa by adding glass water to the sodium hydroxide activator in fly ash-based geopolymer concrete [24]. Songpiriyakij and others stated in their study that the optimum SiO_2/Al_2O_3 ratio required to achieve the highest compressive strength in fly ash-based geopolymer is 15.9 [64].

Factors that significantly affect the mixture properties of fly ash participating in geopolymer concrete are indicated in the form of a chemical and mineralogical combination of fly ash, curing conditions, and molarity of the alkaline solution used in the mixture. The high curing temperature is necessary to increase the effectiveness of fly ash. Because as the curing temperature increases, an increase in the strength of the geopolymer concrete is observed [24, 48, 65, 66].

Figure 2. F Class Fly Ash [71].

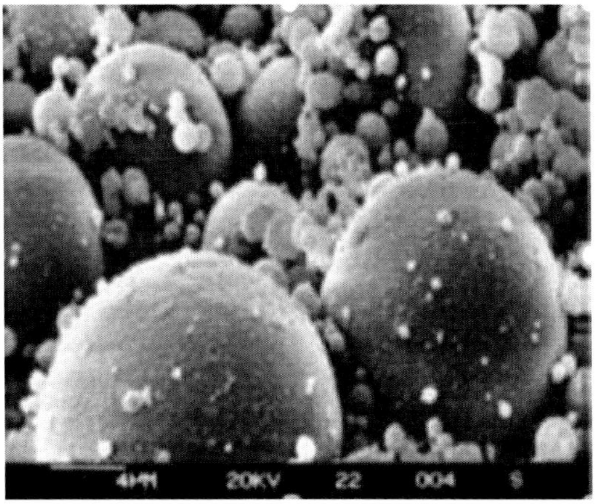

Figure 3. C Class Fly Ash [71].

ASTM standard C618-17a separated fly ash as Class F and C for reference. This classification was based on the calcium oxide content in fly ash. Class C fly ash is produced from high-grade coal, and Class F fly ash is produced from low-grade coal. Therefore, class C fly ash has a higher content of calcium oxide [67, 68]. Fly ash, which is divided into different classes, can create variations in the binding properties of the concrete in which it is joined due to its different chemical content. Fly ash containing lower calcium oxide gave a compressive strength of 65 MPa under high heat-free curing conditions [69, 70]. Thus, F class fly ash is more used in the production of geopolymer concrete. But the lack of binding properties as a result of the failure of F-class fly ash to react at ambient temperature is one of the biggest disadvantages. The reasons why Class C fly ash is not preferred in the production of geopolymer concrete can be explained as having a lower aluminum oxide and silicium oxide content and getting a fast setting time [66].

SEM images of F and C Class fly ash are shown in Figure 2 and Figure 3 [71].

Blast Furnace Slag

Blast-furnace slag is formed by placing a mixture of iron ore and limestone in an oven at a temperature of about 1500°C [72]. BS is chemically composed of calcium oxide, silicate, aluminate, magnesium, and some iron oxide. BS, a pozzolanic material, has the potential to be used as a source material in geopolymer concrete due to the presence of high silica and alumina content in its structure [73]. BS is generally preferred for concrete to gain strength under ambient conditions (heat-free cure) and to increase its physical, and chemical properties and durability [28].

Studies have shown that BS replacement leads to several changes in the mechanical properties of geopolymer concrete.

- In their study, Saladung, Ogawa, and Kawai investigated the effect of BS addition on the mechanical properties of fly ash-based geopolymers. As a result of this study, it was found that adding BS significantly increases compressive strength due to the formation of calcium silicate hydrate (C-S-H) gel, which helps to create a more intense structure [74]. Another study has found that adding BS to fly ash-based geopolymers can improve setting time [75].
- In 2018, Al-Rawi and Tayshi have investigated the effects of steel fiber and BS content on the fresh and hardened properties of self-settling fly ash-based geopolymer concrete. Geopolymer by weight to concrete 0%, 25%, 50%, 75%, and 100% BS has reinforced. They found that the increase in BS hurt the properties of fresh concrete, but significantly increased compressive strength [76].

Studies examining how concrete containing both fly ash and blast-furnace slag react should be considered, as volatile ash alone cannot gain heat-free cure strength, but blast-furnace slag can solve this problem.

For this reason, some studies in which fly ash and blast-furnace slag are used together have been studied and stated as follows:

- In their study, Abhilash and others found that BS and fly ash-based geopolymer concrete mixtures show improved mechanical properties at ambient temperature without the need for heat cure. As in only fly ash-based geopolymer concrete mixtures [77].
- According to the results of Nagalingam and Chokkalingam experiments, it was reported that in all tests, maximum values were obtained for 20% of the replacement of blast-furnace slag with fly ash, and the strength would decrease when it increased to over 20%. It has also been argued that blast-furnace slag and fly ash-based geopolymer concrete are the best alternatives to traditional concrete [78].
- In their review, Vinothkumar and others reported that curing at 70°C gives higher strength than ambient curing. They also concluded that at 48 hours and 70°C heat curing, concrete compressive strength can be increased by adding 12% additional water to the mixture with a water-geopolymer solid ratio of 0.33 [79].
- From conducted research, it has been suggested that the addition of steel fibers to 90% fly ash and 10% blast-furnace slag increases mechanical properties and can be used in site casting work for structures. It was noted that the addition of blast-furnace slag causes the concrete to harden the next day similar to traditional concrete, unlike geopolymer concrete prepared with 100% fly ash, and gains higher compressive strength in a short time [80].
- Fly ash and BS have also been studied to improve the fiber performance of fiber-reinforced geopolymer concrete and have a low demand for water compared to control concrete. A geopolymer concrete, combined with the inclusion of BS in the geopolymer concrete, has been reported to show better results in both ambient curing and heat curing conditions [81].

Silica Fume

Kaolin consists mainly of kaolinite and quartz, while silica fume is completely amorphous silica. Silica fume, known as micro-silica, is a byproduct of the reduction of high purity quartz by coal in electric furnaces in the production of silicium and ferrosilicon alloys [82]. Due to its extreme fineness and high silica content, silica fume is a highly effective pozzolanic material [83]. Silica fume is used in concrete to improve its properties such as compressive strength, adhesion strength, and wear resistance; it reduces permeability and therefore helps to protect the reinforcement steel from corrosion. The addition of silica fume up to 7% significantly increases the geopolymerization process to form a well-defined and compact Matrix. A further increase in the content of silica fume leads to a decrease in the mechanical properties of the reaction product [84]. Interesting high acid resistance of the geopolymer was found to be achieved when using a silica fume activator [41]. It has been determined that silica fume added to the crete as a mineral additive increases pressure and bending strength. Furthermore, if silica fume is to be used, it has been said that adding a superplasticizer to the mixture will increase workability [85].

In one study, the relationship between silica fume and fly ash was investigated. As a result of this study, it was determined that the silica fume ratio of the sample, which gave the best pressure, ben, ding, and tensile strength, was the highest and the fly ash ratio was the lowest [86].

Activators

Chemical activators have a profound effect on the mechanical properties of the geopolymer concrete since the initiation of the polymerization process is under the influence of both the type and concentration of the chemical activator. Al and Si need to be a chemical activator with sufficient concentration to dissolve properly from the raw material. On the other hand, it seems that reactions will be limited if the concentration of the chemical activator is insufficient [61–63].

Alkaline activators such as sodium hydroxide and sodium silicate are often used in geopolymer processes [28].

Strength in geopolymer concretes is directly proportional to the molarity of NaOH, an alkaline activator [87]. In addition, the density and flexural strength of the geopolymer vary according to the NaOH concentration [92]. In his study, Rashad observed that the compressive strength of geopolymer concrete increases to a certain point with the concentration of NaOH solution [37, 38, 88]. But increasing the concentration of NH or the amount of NS has been reported to reduce the workability of geopolymer mortar mixtures [89].

The activation potential of activators with equal concentrations considered in terms of compressive strength is listed as follows [90]:

$Na_2SiO_3 > NaOH > NaOH + Na_2CO_3 > KOH$

Based on ranking, the thaw in NaOH is seen to be higher than in KOH. Despite this, when KOH was used, the average compressive strength of all minerals was 11 MPa. It means 42% higher than NaOH [41].

Alkali-Binding Rate

Another parameter that affects strength is the change in the ratio of alkali binders. The compressive strength of geopolymer concrete added to BS increases with an increase in the alkali-binding ratio [25].

Alumino silicate-based binder and alkali activator are required to synthesize geopolymers (Table 1) [70, 91–93].

Table 1. Principal components of the geopolymer [94]

Aluminosilicate source material	Alkaline activator
Calcined clays, fly ash, rice husk ash, slag, waste glass, copper mine waste, zeolite, pure Al_2O_3-$2SiO_2$ powder and Na_2SiO_3, minerals containing magnesium	NaOH, KOH, $Ca(OH)_2$, NaOH + Na_2CO_3, K_2CO_3 + $Ca(OH)_2$ sodium silicate (other silicates also)

Activator Concentration

Sodium hydroxide (NaOH) concentration is a very important factor for geopolymer synthesis [95]. In studies, it has been claimed that the use of high concentrations of sodium hydroxide (NaOH) in the geopolymer increases compressive strength [36], [96]. Yilmazer P. and others in their study have determined the optimal concentration of sodium hydroxide for geopolymer concrete at 12 mol [97].

Memon and others investigated the compressive strength effect of sodium hydroxide (NaOH) concentration on fly ash-based self-settling geopolymer concrete. They examined the compressive strength of samples with different NaOH concentration values. The results show that the NaOH concentration of the sample, which gives the best compressive strength, is 12 mol [98].

Wang, Li, and Yan have studied the synthesis and mechanical properties of metakaolinite-based geopolymers. In their work, metakaolinite-based geopolymers were synthesized at 20°C under NaOH solution (4-12 mol/L) and sodium silicate activation. The results show an increase in the bending and compressive strength of geopolymers by increasing the NaOH concentration by 4-12 mol/L [99].

Kumaravel and Girija have determined the compressive strength values of geopolymer concrete samples prepared using 4 different NaOH concentrations. In their studies, geopolymer concrete samples were prepared with the highest compressive strength value of 12 M [100].

Fibers

Concrete is one of the most widely used building materials in the world. But because the final tensile stress of the concrete is very small, the tensile strength and impact resistance is very low compared to the compressive strength. In a study, it was reported that the tensile strength of concrete is about 10% of the compressive strength [101]. For this reason, materials should be used to prevent this sudden breakage by increasing ductility in the concrete. Some of these materials that can be reinforced into concrete are fibers of different properties [102]. At this point, the inclusion of different types of fibers in the geopolymer Matrix proved very useful in achieving the desired mechanical and durability properties [103]. The presence of fibers allows the density of the matrix to increase. Thus, the geopolymer causes a decrease in concrete's specific density and water absorption. These effects apply to all fibers and vary depending on fiber content [37].

The mechanical strength and modulus of the fibers decrease as the fiber diameter increases. When the fiber content increases at a critical concentration, the fibers tend to disperse erratically, forming clusters or balls. Even very fluid matrices may not be able to pass properly through the congested fiber network. Therefore, the fresh Matrix requires more vibration to lose its hard static mode and form the mold. Fiber properties affect the strength development of composites over time, for example, increasing fiber content regardless of fiber type and geometry reduces the fluidity of fiber-reinforced geopolymer composites [107]. This can be measured in the form of a decrease in the traditional collapse value (ASTM c143), a fluidity test

(ASTM C230), an increase in the VeBe time test, a time to pass through the V-funnel (EFNARC-2002), or an increase in flow stress and viscosity measured using a rheometer. The reduced workability can be explained as a result of the increase in flow stress of the fresh composite due to the contact network between the hard fibers within the Matrix. As a result, the content of the fibers increases with an increase in the equivalent width and aspect ratio. Although fibers shouldn't reduce workability, increasing fiber content causes many micro-cracks to occur, rather than a few large macro cracks [108]. One of the most important conditions that must be achieved in all-fiber reinforced concrete is that the fiber distribution is homogeneous, which prevents cracks formed in the concrete and slows its progress, increasing the durability of the concrete. Another condition is that this distribution does not deteriorate after mixing concrete [104, 109, 110]. Fibers that are randomly distributed in the concrete can bond with each other and stop the expansion of the crack. In this way, adding fiber to normal concrete can increase the mechanical behavior of concrete [111]. In addition, in other studies, it has been emphasized that fibers improve their properties such as toughness, fatigue, and initial crack strength in concrete [112, 113].

When used at a recommended dosage, fibers can increase the resistance to tensile stresses [114].

Many studies have been conducted to investigate the effect of fiber used in concretes. In these studies with various fibers, fiber use generally gave positive results because adding fiber to concrete can increase pressure, tensile and shear strength, bending resistance, as well as impact resistance [115, 116].

- In the Yap, Alengaram, and Jumaat studies, they found that adding polypropylene and nylon fibers can increase the mechanical properties of concrete [117]. In another study, it was emphasized that polypropylene fiber gave positive results in preventing crack formation, but its effect on the compressive strength of concrete was almost minimal [118].
- In Ma and Tan's studies, they examined the mechanical properties and durability of ceramic fiber-reinforced cemented composites. In this study, they reinforced ceramic fiber to cement mortar by 5% volume. The results show that ceramic fiber reinforcement can increase bending strength by 40%, and when silica fume is added, the ratio increases to 100% [119].

Figure 4. Some fibers are used in geopolymer concrete (respectively; carbon, basalt, kevlar, polypropylene, and steel fiber) [104]–[106].

- Bhalchandra and Bhosle investigated the effect of glass fiber reinforcement on density, flexural strength, and compressive strength on geopolymer concrete composites. The results show that glass fiber reinforced geopolymer concrete composites show higher strength compared to fiber-free geopolymer concrete and OPCC [120].

Carbon Fibers

One of the biggest disadvantages of geopolymer concretes is brittleness because the brittleness of concrete increases as the strength increases [121]. To fix this problem, it is necessary to use a material that increases ductility. It is thought that the carbon fibers involved in concrete will have a beneficial effect on compressive strength, flexural strength, and direct tensile strength, where the mode of fracture varies significantly from brittle to ductility [122].

Carbon fiber was first discovered in Cleveland, Ohio in 1958. In the early days, the purpose of use was only isolation, filtration materials, and illumination. But it was later discovered that when used with epoxy resin, it forms composite structures with high strength [105].

Structure and Features

Carbon fibers are mainly based on sources such as polyacrylonitrile (PAN), oil field, and rayon fibers. The performance of carbon fibers is mainly determined by the structure of graphite crystallites in their microstructure [105, 108]. Carbon fibers are known as fibers with the highest specific strength and lightness among all reinforcement fibers. But it reduces the workability of concrete [122]. The fibers show outstanding tensile strength and modulus at high temperatures, very low thermal expansion, chemical, and thermal stability, and high electrical & thermal conductivity compared to other materials. In addition, carbon fibers are perfectly elastic, which makes them less affected by fatigue deformation during loading and unloading cycles [108].

Despite the high mechanical strength, thermal resistance, and energy absorption achieved for carbon fiber reinforced geopolymers, the high cost and partial oxidation of carbon at high temperatures limit their practical application [108, 123].

Some studies on silica fume and carbon fiber reinforced concrete are listed as follows:

- In the study of Wen and others, it was observed that silica fume reinforced to the concrete mixture has a strength-enhancing effect in concrete with carbon fiber addition. When the causes were examined, it was found that silica fume additive helped to precipitate the fiber. In addition, it has been found that it fills gaps caused by fiber due to its fine grain structure texture [124].
- Due to the air gap caused by the addition of carbon fibers, the ultrasonic permeability (USG) speeds of fibrous concrete have decreased. Due to the gap-filling effect of silica fume, concrete with silica fume addition has been argued to have a higher USG rate than concrete without silica fume, and it has been proven by experiments that carbon fibers do not have a noticeable contribution to strength despite silica fume [125].

Basalt Fibers

In other words, the toughness present in fiber-reinforced concrete resists impact, crack, and break thanks to its energy damping property. This feature is one of the most important reasons for adding fiber to concrete [104, 109, 110]. For this reason, the structure and properties of basalt fibers were studied.

Structure and Features

Basalt fibers increase the impact resistance and cleavage resistance of concrete. After cracks form in the mortar, it delays the expansion of the crack by bridging effect. In this way, basalt fibers allow excessive amounts of energy to be absorbed [126]. In addition to all these, basalt fibers have high durability, temperature resistance, and resistance. Thus, it can be an alternative to materials whose chemical structure is not active under extreme temperatures [127]. The reason it shows such good properties is because it is obtained from volcanic rocks. Basalt fibers are formed by melting basalt at temperatures of 1300-1700°C, dividing it into very thin pieces, and obtaining it in fiber form [128, 129]. No other additional additives are used during fiber production [136]. For this reason, basalt fiber production can be low-cost [130].

This fiber is a typical ceramic fiber, it is easy to disperse in the mixture when mixed with cement mortar and concrete. In general, concrete reinforced with basalt fiber has good workability, volume stability, and sealing while fresh. While hardened, it has good properties such as high impact, crack, and thermal resistance [126].

Xu and W. Li investigated the effect characterization of basalt fiber reinforced geopolymer concrete in their studies. As a result of this research, they found that basalt fiber reinforcement can significantly improve the deformation and energy absorption properties of geopolymer concrete, but there is no noticeable difference in dynamic compressive strength [131].

Kevlar Fibers

Poly-aramid fiber, known as Kevlar among synthetic fibers, is a material consisting of solid fibers of very lightweight carbon origin. It has many good mechanical properties [132]. Compared to other synthetic fibers, it has significantly lower fiber elongation and higher tensile strength and modulus [132–136].

Structure and Features

Kevlar fibers are a low-density insulating material. Its structure resists up to approximately 500°C, so it is also resistant to fire. Modulus of elasticity, tensile, impact, fatigue, and friction strength is high. Kevlar fiber has higher durability and weight ratio compared to steel, but steel is superior to Kevlar fiber in durability and volume [137]. Because of these superior mechanical properties, they are widely used in industrial and military applications [138, 139]. Kevlar fibers are classified as K29, K49, K100, K119, K129, and KM2 [137].

In a research study, the nature of the K49 fiber under different loading conditions was investigated. The effect of axial compression on recoverable stress properties was found to be very small [140].

In another study, it was found that the dynamic properties of K49 fiber, such as modulus of elasticity, tensile strength, maximum stress, and toughness, depending on the tensile ratio [141].

Steel Fibers

Brittleness can be removed by using steel fibers [142–144]. It increases the strength of the concrete it is added to. However, the amount of addition is affected by various factors such as dosage, size, shape, distribution, and casting director [145].

The length of the steel fiber is an important parameter that increases the efficiency of the fibers in their bending behavior [145–147]. However, bending performance is also affected by fiber orientation and distribution, which is strongly dependent on the length and aspect ratio of the steel fiber [146].

Structure and Features

The tensile stress of steel fibers used in concrete is about approximately 1200 MPa. Elastic limits are lower than 0.2%. The diameters of the fibers are between 0.13 and 1.0 mm. Their lengths vary between 13 mm and 70 mm. The volume of steel fiber in concrete varies between 0.5% and 3% [148].

Çelikten et al. produced blends without waste steel fiber and blends containing 1% and 2% waste fiber by volume. And the effects of fiber content on the mechanical properties of geopolymer mortars were investigated. As a result, they reported that steel fibers obtained from waste tires improved the mechanical properties of alkali-activated blast furnace slag mortars.

Polypropylene Fibers

Polypropylene, the most common commercial thermoplastic polymer, produced 45 million tons worldwide in 2001, creating a market of 65 billion dollars. Especially isotactic polypropylene (iPP) has a very high power to weight ratio thanks to its low density and high strength. For this reason, it is the most widely used polymer in every field from automotive parts to the textile and packaging industry.

Polypropylene is generally produced in 3 different ways. It is mostly seen in the market as homopolymer, copolymer, or block copolymer. The co-monomer used in the production of copolymers is usually ethylene and ethylene-propylene rubber is obtained.

Structure and Features

Polypropylene crystals have a melting point of around 160 °C and are usually processed above 200°C. Polypropylene is a polymer sensitive to UV rays from the sun and high temperatures during heat treatment.

Yadollahi and Güneş examined the effect of polypropylene fibers on the mechanical behavior of geopolymers in their study in 2018. At the end of the experiments, they observed that compressive strength and flexural strength increased with the addition of polypropylene fibers for all mixtures and samples.

As a result, some mechanical properties of the fibers studied are given in Table 2.

Table 2. Mechanical properties of fibers [149, 150]

Fibers	Tensile strength (GPa)	Density (g/cm^3)	Elasticity Modulus (GPa)
K29	3.60	1.44	65.5 [139]
K49	3.80	1.44	131
K129	3.40	1.47	186
Carbon fiber (small)	6.21	1.80	297
Carbon fiber (medium)	5.10	1.80	241
Carbon fiber (large)	3.62	1.74	228
Basalt Fiber	2.9-3.1	2.65	85-87
Steel Fiber	2.76-27.60	7.80	200
Polypropylene	5.52-7.59	0.90	3.6

Curing Conditions

Geopolymer samples are heat-treated at a constant temperature of 350°C with pressure. Compressive strength increases in hot press curing. At higher pressures, the trapped air pores in the geopolymer matrix are removed, which leads to an increase in density [94, 151].

In their study, Joseph and Mathew reported that increasing the curing temperature to 100°C would increase the compressive strength in geopolymer concrete, but temperatures above 100 °C would lead to a decrease in compressive strength [18].

Cracks due to internal stresses will occur if the sample overheats during curing. This situation weakens the structure and mechanical strength decreases significantly [152, 153]. Also, a lower temperature (75°C) for a longer period (72 hours) of curing, leads to the formation of microcracks [94, 154].

It has been stated that the performance of geopolymer concrete will increase by up to 70% by increasing the curing temperature [8]. It has been found that the curing temperature affects the strength of the metakaolin-based geopolymer [8]. At the same time, Singh (2016), in his study of geopolymer concretes, reported that the compressive strength of geopolymer cubes increases as the curing time and temperature increase [25]. Vinothkumar and others (2019) found that curing at 70 °C gives higher strength than ambient curing. They also concluded that concrete compressive strength can be increased by adding 12% additional water to the mixture with a water-geopolymer solid ratio of 0.33 at 48 hours and 70°C thermal curings [79].

In studies on the curing temperature of fly ash-based geopolymer concrete, it was observed that 70°C cured geopolymer concrete samples give the highest compressive strength [155, 156].

Another study with geopolymer samples had less strength than furnace-cured samples. However, a 75/25 fly ash/blast-furnace slag ratio has been reported to give maximum strength in both cured samples. Blast-furnace slag dosage beyond 25% has been shown to slightly reduce strength in mixture designs. In conclusion, ambient-cured samples have the highest average compressive based on fly ash and blast-furnace slag showed ambient-cured strength, a higher density, and a Poisson ratio than the furnace-cured sample [157].

In studies with blast-furnace slag and rice husk ash, it was emphasized that the curing temperature should be 70°C for higher strength [158, 159].

Al-Shether, Al-Attar, and A. Hassan used different methods to investigate the curing of metakaolin-based geopolymer concrete. To study the curing behavior of geopolymer concrete, they conducted experiments under direct sunlight and in the laboratory, both in summer and winter. An average pressure strength of metakaolin-based geopolymer concrete during the summer season of 7 days was found to be 28.39 MPa under sunlight and in a laboratory ambient. This value was measured at an average of 18.44 MPa during the winter season. According to the results, metakaolin-based geopolymer concrete gave the best compressive strength in summer and under direct sunlight [160].

Durability in Geopolymer Concrete

Compressive Strength

In 1972, under the influence of kaolinite clay and alkaline activators, the mortar compressed with a pressure of 15 MPa was pressed on fiberboard at 100-130ºC with a pressure of 30 bar steam press. As a result, fireproof wood panel production was carried out. The characteristic compressive strength of the panel is 80 MPa [9].

Studies using various mineral additives are available in geopolymer concrete. These studies show that there are changes in compressive strength according to the mixture principles of the geopolymer.

- Chandra pacemaker and Sarath Chandra Kumar have tried to achieve the best compressive strength by using different proportions of blast-furnace slag and metakaolin as pozzolanic material. The results show that the percentage of metakaolin is reduced and replaced by blast-furnace slag substitution, which leads to a decrease in compressive strength [161].
- Celik and others tried to capture the best compressive strength by mixing metakaolin and colemanite in different proportions. In this study, they examined the compressive strength and flexural strength of 5 different samples. The results show that the sample with 90% metakaolin and 10% colemanite gives the best compressive strength and flexural strength [162].
- Aygörmez and others studied silica fume, metakaolin, and colemanite. Compressive strength and flexural strength of the geopolymer composite created using different mixture ratios were examined. This study showed that the use of silica fume reduces the

gap rates of geopolymer matrices and thus increases compressive strength & bending strength. But excessive use of silica fume has shown negative effects on compressive strength and flexural strength [163].

- Keerthy and Kumar tried to capture the best compressive strength in their study using different proportions of blast-furnace slag and fly ash. They conducted this study separately for 8 mol and 10 mol. The results show that reducing the blast-furnace slag ratio and replacing fly ash leads to a decrease in compressive strength, and a concentration of 10 moles compared to 8 moles gives better results [164].

Impact

An impact resistance reinforced geopolymer concrete can be used in airport runway concretes, concrete roads, and security barriers. It can be used in strategic structures to minimize the impact of sudden loads caused by rocket attacks, etc.

This impact resistance can be measured in 4 different ways:

1. Moving Pendulum (Charpy, Izod)
2. Structural elements dropped from a certain height
3. Explosive materials
4. Falling Ball
    - Decline in fixed elevation
    - Decrease in variable elevation [165].

But the modern understanding of concrete's impact resistance is limited. The basis of the problem is the lack of a standard testing technique for testing concrete under impact load [166].

Impact Resistance

Dynamic loads (impact) are often an attractive area for researchers to characterize the response of material exposed to such loads, especially in military applications. Damage caused by impact loads varies depending on many factors, such as impact speed, the geometric properties of the bullet, the material properties of both the bullet and the structure, and the geometric properties of the target (thickness) [167].

Many researchers have studied the behavior of OPCC-based material under impact loads and how to improve this behavior [167–171]. However, limited research in the literature has focused on the reaction of alkali-activated materials and their resistance to impact loads. Previous studies have indicated the weak performance of geopolymers under impact load. For this reason, it has been proposed to include fiber to increase impact resistance by researchers working on the subject in the literature [131, 172].

Studies on the impact effect on geopolymer concrete have been compiled as follows:

- The behavior of reinforced geopolymer concrete slab under repeated low-speed impact load was investigated by Madheswaran et al. (2007). OPCC and geopolymer concrete slabs measuring 1000x1000x60 mm were poured to study their behavior under impact load. The slabs are mounted on a support frame with hinged support conditions. An 8.4 kg hammer was used for impact testing. A dynamic load cell is connected to the tape recorder via a strain gauge instrument and an amplifier. The TEAC instrumentation recorder was used to record impact load, deformation, and acceleration signals for each drop. From test results, it was found that geopolymer concrete plaques performed better than normal portland cemented concrete plaques during the initial cracking and failure stages under impact testing. The results were compared using the ANSYS program. Flat slabs failed by puncture, while fiber-reinforced slabs showed only surface (cracking) failure. Progressive deterioration in concrete hardness has occurred under low-energy repeated impact load [173].
- In experimental studies by Nagan and Mohan (2014), the resistance of geopolymer mortar slabs under impact load by dropping a steel ball from a significant height was found. For this study, samples of 230x230x25 mm sizes were used with different combinations of chicken mesh and rectangular welding mesh and subjected to impact load by drop weight test. Then, the impact energy required for the first crack and the last crack was calculated. The combination of chicken mesh and rectangular welding mesh was found to perform better in the case of energy absorption and residual impact strength. Compressive strength, flexural strength, and split tensile strength of 10 m geopolymer mortar samples were found to be 36.05%, 33%, and

27.7% higher compared to cement mortar samples, respectively. Furthermore, with the increase in reinforcement volume, the energy absorption of the geopolymer ferrocement sample was reported to increase compared to the cement mortar sample [174].

- In this experimental study, the behavior of ferrocement slabs under impact load was studied by changing parameters such as the number of wire mesh layers, SBR polymer content, and drop height. For this study, 500x500x50 mm samples with a drop weight of about 1.3 kg, and a drop height of 2.5 m, 1.2 m, and 0.83 m were used. The required blows for the first and last crack have been recorded. The impact resistance capacity of the samples has been calculated. They concluded that the impact energy absorption capacity of the ferrocement sample increased as the percentage of polymer content and the number of wire mesh layers increased [175].
- It was found that the energy absorbed by the material during impact depends on the energy distribution or the deformation rate of the energy [176].
- Abubakr and others conducted a falling weight experiment to calculate impact strength. As a result, they notcd that alkali-activated slag exhibits 70% and 40% better compression and tensile strength, respectively, compared to OPCC. They also reported that the deformation and energy absorption of AAS as a result of impact loading can vary with the activator used and that AAS mixtures show better mechanical properties compared to cement-based mixtures in terms of compression, tensile, and impact absorption. They concluded that compressive strength and impact energy absorption is increased by increasing silica modulus and sodium oxide percentage in general [177].

In the literature studies studied, it was stated that geopolymer concretes are more durable under sudden loading than OPCC. In these studies, geopolymer has been studied both physically and chemically.

Temperature, Freeze-Thaw, and Explosion Resistance in Geopolymer Concrete

The explosions are causing public concern. The concrete breaks down as a result of the explosion. This causes a shrapnel effect. The shrapnel effect increases the loss of life and property. Several studies are being carried out to reduce these negative effects of explosions. For this reason, the investigation of the reaction of geopolymer concrete to explosion loading has gained importance in recent years.

To understand the reaction of concrete against explosion loading, it is necessary to first know the properties of explosives. Explosives can be in solid-liquid and gaseous forms, and the rate of explosion varies according to these physical states. Explosives cause more damage in solid form than in other forms. Shock waves from an explosive that explodes in a solid state can reach a temperature of 3000-40000°C and a pressure value of about 30000 MPa [178–180]. This high temperature and pressure value caused by explosions can cause heavy damage to concrete. Studies have shown that geopolymer concrete gives a higher performance compared to OPCC at high temperatures. In general, the temperature increase rate decreases after reaching an estimated temperature of 800°C in the first 30 minutes under the influence of fire. It reaches a constant temperature between 1000°C and 1100°C in about 2.0-2.5 hours. Serious damage and spills can occur if the OPCC is constantly exposed to such extreme temperatures. Damage can change the balance of the reinforcement, resulting in the collapse of the structure. This can both increase economic costs and occur a significant risk to human life. For this reason, various studies are carried out on different types of concrete against high temperatures. One of these types of concrete is geopolymer concrete. In some studies related to the resistance of geopolymer concrete to high temperature in the literature, it has been found that the compressive strength of geopolymer concrete cylinders exposed to fire at 800°C has increased [163, 181].

Another study showed that samples of foamed porous fly ash-based geopolymer paste increased compressive strength after exposure to temperature (up to 1000°C) [182].

In addition, samples can be exposed to high pressures and high temperatures in explosion experiments. Explosives must be positioned at an accurate distance for the concrete to not experience heavy destruction and remain in a state that can be examined. The mass of the explosive is directly proportional to the distance. The effective distance (Z) for explosion experiments is shown in Equation (1) [183].

$$Z = R/\sqrt[3]{W} \tag{1}$$

In this formula, R is the distance away (m) and W is the explosive mass (kg).

- In explosion experiments, geopolymer concrete was usually reinforced with fiber. In their study, Li and others found that steel wire mesh and fibers can effectively improve the structural behavior of concrete elements against explosion load in both experimental and numerical studies [195]. In their study, Meng and others examined the behavior of steel fiber reinforced geopolymer concrete panels against explosion loading and observed that steel mesh reinforced geopolymer concrete panels better dissipate explosion energy [184].
- In another study, Meng and others examined methane gas explosions (flare-ups) in buried service tunnels. They designed the buried service tunnel with steel fiber geopolymer concrete composite. The results after the explosion are as follows, geopolymer concrete has been found to have good compressive and flexural strength under static load. In addition, the compressive strength of fiber-free geopolymer concrete was found to be 61 MPa, and the compressive strength of steel fiber reinforced geopolymer concrete was found to be 74 MPa [197]. In another study, steel wire braided geopolymer concrete was subjected to a TNT explosion. The geopolymer is made of concrete, silica fume, and blast-furnace slag. As a control sample, C30 concrete slabs reinforced with steel reinforcement were tested. As a result of the TNT explosion, geopolymer concrete slabs reinforced with steel wire mesh provided a proper strain distribution. In other words, it was found that geopolymer concrete performed better structural performance against traditional C30 concrete slabs under the same explosion loads [185].
- In another study, Meng and others investigated a new type of building element made of high-performance alkali-activated geopolymer concrete reinforced with steel wire mesh subjected to a 50 kg TNT explosion. In this study, steel-reinforced C30 concrete slabs were produced to compare with geopolymer concrete samples. Both the C30 concrete slab and the steel wire reinforced geopolymer concrete slab did not give positive results in the explosive positioned at a distance of 3 meters. But in explosion experiments conducted at 5 meters and 7 meters, the geopolymer samples showed less crack formation than the C30 concrete samples [186].

- Another important parameter in geopolymer concrete is freeze-thaw resistance. Freezing-thaw resistance is associated with the size of the pores of concrete [187]. During freezing-thawing, the solution inside the pores freezes and turns into ice, thus increasing the internal tension [188, 189]. If this stress value exceeds the strength of the concrete, it can create cracks in the concrete [190].
- Studies are available in the literature to reveal the effect of freeze-thaw resistance on geopolymer concrete. In their study, Zhao and others examined the freeze-thaw resistance of fly ash-based geopolymer concrete with different proportions of slag reinforced. By increasing the slag ratio from 10% to 50%, the freeze-thaw resistance increased from 5 to 225 freeze-thaw cycles [191].
- Yuan and others studied the effects of different fiber additives on the freeze-thaw resistance of fly ash-based geopolymer concrete. Three different fibers were used in the study. These are polypropylene, polyvinyl alcohol, and steel fibers. These fibers added to the geopolymer concrete were unable to prevent the onset of micro-cracks. But these three fibers used in geopolymer concrete prevented the cracks from spreading [192].

Acid Effect in Geopolymer Concrete

Acids are an important parameter that affects the strength of concrete. For this reason, many studies have been conducted on the acid resistance of concrete. In general, the type of acid that is the subject of studies is sulfuric acid. Because it is exposed to sulfuric acid in mining, sewage, and food processing ındustries applications [193].

Kumaravel and Girija tested geopolymer concretes with different NaOH concentrations in their work. Later, they investigated the resistance of these samples to sulfuric acid. Compressive strength was obtained as 24.20, 33.50, 25.60, and 28 N/mm^2, respectively, in control samples with a concentration of 14 M, 12 M, 10 M, and 8 M NaOH. In this study, samples were then immersed in H_2SO_4 solutions of different concentrations and monitored for 90 days. The results are as follows:

Table 3. Strength Differences for Different Concentrations and Different Mole Values

H_2SO_4 concentration→	0.5%	1%	2%
Moles ↓	Decrease in strength %		
8	4	9	19.1
10	5.35	13.27	25.5
12	3.70	8.59	16.72
14	4.76	10.5	26.70

The results from here show that the sample with the best values of 12 M provides [100].

Ariffin and others, on the other hand, produced blended ash geopolymer concrete using powdered fuel ash and palm oil fuel ash in their work. The OPCC and blended ash-based geopolymer concrete and control sample was exposed to a 2% solution of sulfuric acid for up to 18 months. Blended ash geopolymer concrete suffered a 7.3% loss of compressive strength in the first month it was exposed. It then achieved a loss of 1.6% each month and a total loss of 35% in 18 months. OPCC samples suffered a great loss of strength in the first month they were exposed and reached a total loss of 68% of strength in 18 months [194].

As a result of their research, Wallah and Rangan concluded that geopolymer concrete based on low calcium fly ash showed good acid resistance under sulfate attacks [62].

Bakharev studied the behavior of geopolymer materials produced using F-class fly ash and alkali activators in the sulfate ambient. Geopolymer materials were exposed to 5% sodium sulfate (Na_2SO_4), magnesium sulfate ($MgSO_4$), and sodium sulfate (Na_2SO_4) + magnesium sulfate ($MgSO_4$) solutions over 5 months. The samples used in the study were prepared using sodium silicate, sodium hydroxide, and potassium hydroxide. The control sample in the study is OPCC. A sample prepared using only sodium hydroxide in a solution of sodium sulfate showed the best result, and there was some power fluctuation in geopolymer samples immersed in solutions of magnesium sulfate and sodium sulfate + magnesium sulfate [195].

Ganesan, Abraham, and Raj examined the durability properties of steel fiber reinforced geopolymer concrete. They observed performance changes of prepared samples against acid and sulfate attacks. The prepared geopolymer samples are fly ash-based and 5 pieces. Steel fiber ratios of geopolymer samples differ from each other (by Volume 0%, 0.25%, 0.5%, 0.75% and

1.00%). Two different OPCC samples with 0% and 0.5% steel fiber by volume were used as the control sample. 6 samples were produced from each mixture. 3 samples were immersed in a 3% solution of sulfuric acid (H_2SO_4) for 180 days. The remaining 3 samples were immersed in a 3% sodium sulfate (Na_2SO_4) solution for 180 days. In geopolymer concrete samples, the loss of compressive strength caused by acid and sulfate attack is around 20%, while in OPCC samples, this value is around 35%. Geopolymer samples were less damaged by acid and sulfate attacks [196].

Conclusion

This study is a general summary of the studies on the physical and mechanical properties of geopolymer concrete. As a result of the research and the results obtained, the potential of geopolymer concrete to be used as a construction material is high. According to the OPCC, it's more environmentally friendly which places geopolymer concrete one step ahead.

Geopolymer Concrete:

- It performed better against high temperatures than OPCC.
- It showed good results against explosion loading when various fibers are added.
- Approached ultimate compressive strength in a short curing time of 7 days.
- It showed high compressive strength when prepared with the right mixtures.
- It performs well against acid reactions.
- When prepared with the right mixtures, geopolymer concrete has a lower clearance rate than OPCC.
- When prepared with the right mixtures, the Poissonsson ratio and elasticity modulus of the geopolymer concretes are higher than the OPCC.

Different mixtures have been observed to affect the mechanical properties of geopolymer concrete. But to understand its full effect on mechanical properties, it is necessary to try different variations in mixtures. For this reason, it is necessary to investigate more geopolymer concrete and expose its unknowns.

References

[1] Saka R. C. and S. Subaşi, "Sepiyolit İkameli Çimentolu Karışımların Mekanik ve Fiziksel Özelliklerinin İncelenmesi," *El-Cezeri Fen ve Mühendislik Derg.*[« Investigation of Mechanical and Physical Properties of Sepiolite Substituted Cement Mixtures," *El-Cezeri Journal of Science and Engineering*], vol. 5, no. 2, pp. 681–692, May 2018, doi: 10.31202/ecjse.364622.

[2] McCaffrey, R. "*GCL: ENVIRONMENTAL SPECIAL ISSUE 2002 CLIMATE CHANGE AND THE CEMENT INDUSTRY*," 2002.

[3] Malhotra, V. M. "Introduction: Sustainable Development and Concrete Technology," *Concr. Int.*, vol. 24, no. 7, Jul. 2002.

[4] Naik, T. R. "Sustainability of Concrete Construction," *Pract. Period. Struct. Des. Constr.*, vol. 13, no. 2, pp. 98–103, May 2008, DOI: 10.1061/(ASCE)1084-0680(2008)13:2(98).

[5] Hardjito D. and B. V. Rangan, "*Development and properties of low-calcium fly ash-based geopolymer concrete*," 2005.

[6] Vora P. R. and U. V. Dave, "Parametric studies on compressive strength of geopolymer concrete," in *Procedia Engineering*, Jan. 2013, vol. 51, pp. 210–219, DOI: 10.1016/j.proeng.2013.01.030.

[7] Turner L. K. and F. G. Collins, "Carbon dioxide equivalent (CO2-e) emissions: A comparison between geopolymer and OPC cement concrete," *Constr. Build. Mater.*, vol. 43, pp. 125–130, 2013, DOI: 10.1016/j.conbuildmat.2013.01.023.

[8] Ali, A., J. T. Nidhin, and B. Parappattu, "Geopolymer As Repair Material -a Review," *Int. J. Innov. Res. Adv. Eng.*, vol. 09, no. 3, pp. 2349–2763, 2016.

[9] Davidovits, J. "Properties of Geopolymer Cements," *First Int. Conf. Alkaline Cem. Concr.*, pp. 131–149, 1994, Accessed: Dec. 03, 2020. [Online]. Available: www.geopolymer.org.

[10] Habert, G., J. B. D'Espinose De Lacaillerie, and N. Roussel, "An environmental evaluation of geopolymer based concrete production: Reviewing current research trends," *J. Clean. Prod.*, vol. 19, no. 11, pp. 1229–1238, Jul. 2011, DOI: 10.1016/j.jclepro.2011.03.012.

[11] Gomes, K. C., M. Carvalho, D. de P. Diniz, R. de C. C. Abrantes, M. A. Branco, and P. R. O. de Carvalho, "Carbon emissions associated with two types of foundations: CP-II Portland cement-based composite vs. geopolymer concrete," *Rev. Mater.*, vol. 24, no. 4, 2019, DOI: 10.1590/s1517-707620190004.0850.

[12] Živica, V., M. T. Palou, and M. Križma, "Geopolymer Cements and Their Properties: A Review," *Build. Res. J.*, vol. 61, no. 2, pp. 85–100, 2015, DOI: 10.2478/brj-2014-0007.

[13] Ma, C. K., A. Z. Awang, and W. Omar, "Structural and material performance of geopolymer concrete: A review," *Constr. Build. Mater.*, vol. 186, pp. 90–102, 2018, DOI: 10.1016/j.conbuildmat.2018.07.111.

[14] Davidovits, J. "Soft Mineralurgy and Geopolymers," *Geopolymer 1988*, vol. 1, no. June 1982, pp. 19–23, 1988, Accessed: Feb. 25, 2021. [Online]. Available: https://www.researchgate.net/publication/328783506.

[15] Pan Z. et al., "Damping and microstructure of fly ash-based geopolymers," *J. Mater. Sci.*, vol. 48, no. 8, pp. 3128–3137, Apr. 2013, DOI: 10.1007/s10853-012-7090-y.

[16] Pan, Z., Z. Tao, Y. F. Cao, R. Wuhrer, and T. Murphy, "Compressive strength and microstructure of alkali-activated fly ash/slag binders at high temperature," *Cem. Concr. Compos.*, vol. 86, pp. 9–18, Feb. 2018, DOI: 10.1016/j.cemconcomp. 2017.09.011.

[17] Park, J. W., K. Y. Ann, and C. G. Cho, "Resistance of Alkali-Activated Slag Concrete to Chloride-Induced Corrosion," *Adv. Mater. Sci. Eng.*, vol. 2015, 2015, DOI: 10.1155/2015/273101.

[18] Joseph B. and G. Mathew, "Influence of aggregate content on the behavior of fly ash-based geopolymer concrete," *Sci. Iran.*, vol. 19, no. 5, pp. 1188–1194, Oct. 2012, doi: 10.1016/j.scient.2012.07.006.

[19] Şinik, O. "*Geopolimer Betonlarda Dayanıklılık Özelliklerinin Araştırılması,*[*Investigation of Durability Properties in Geopolymer Concretes*]" 2019.

[20] Davidovits, J. "*Pyramids of Egypt Made of Man-Made Stone, Myth or Fact?*" 1984.

[21] Davidovits, J. "Geopolymer chemistry and properties," *Geopolymer*, vol. 88, no. 1, pp. 25–48, 1988.

[22] Davidovits, J. "Who built the pyramids? " *Harv. Mag.*, vol. Jul, pp. 42-49,99, 2003, Accessed: Dec. 25, 2020. [Online]. Available: Davidovits J, Geopolymer chemistry and applications. Institut Géopolymère, Saint-Quentin, France, 2008, ISBN 2-86553-190-2 2005.

[23] Van Jaarsveld, J. G. S., J. S. J. Van Deventer, and A. Schwartzman, "The potential use of geopolymeric materials to immobilise toxic metals: Part II. Material and leaching characteristics," *Miner. Eng.*, vol. 12, no. 1, pp. 75–91, 1999, DOI: 10.1016/s0892-6875(98)00121-6.

[24] Palomo, A., M. W. Grutzeck, and M. T. Blanco, "Alkali-activated fly ashes: A cement for the future," *Cem. Concr. Res.*, vol. 29, no. 8, pp. 1323–1329, Aug. 1999, DOI: 10.1016/S0008-8846(98)00243-9.

[25] Singh, A. K. "Strength and Durability Test of Fly Ash and GGBS Based Geopolymer Concrete Compressive Strength in MPa Strength Mpa Alkaline / Binder Ratio," *International J. Eng. Res. Appl.*, vol. 6, no. 8, pp. 139–142, 2016.

[26] Davidovits, J. "Geopolymers - Inorganic polymeric new materials," *J. Therm. Anal.*, vol. 37, no. 8, pp. 1633–1656, Aug. 1991, DOI: 10.1007/BF01912193.

[27] Davidovits, P. J. "30 Years of Successes and Failures in Geopolymer Applications. Market Trends and Potential Breakthroughs .," *Geopolymer 2002 Conf.*, pp. 1–16, 2002, Accessed: Mar. 30, 2021. [Online]. Available: www.geopolymer.org.

[28] Yalghuz, M. R. "*Farklı tip malzemeler kullanılarak geopolimer harç üretimi ve dayanıklılığının incelenmesi*,[*Investigation of geopolymer mortar production and durability using different types of materials*]" 2020.

[29] Aleem M. I. A. and P. D. Arumairaj, "*Geopolymer concrete- a review,*" 2012.

[30] Rangan, B. V. "Fly Ash-Based Geopolymer Concrete," *Geopolymer Cem. Concr.*, vol. 7982, no. December, pp. 68–106, 2010, Accessed: Dec. 03, 2020. [Online]. Available: https://www.researchgate.net/publication/230717147.

[31] Pacheco-Torgal, F., J. A. Labrincha, C. Leonelli, A. Palomo, and P. Chindaprasirt, *Handbook of Alkali-Activated Cements, Mortars, and Concretes*. 2014.

[32] Davidovits J. and D. Comrie, "Long term durability of hazardous toxic and nuclear waste disposals," *1st Eur. Conf. Soft Miner. Compiegne, Fr.*, vol. 1, pp. 125–134, 1988, Accessed: Mar. 31, 2021. [Online]. Available: https://www.researchgate.net/publication/312970236.

[33] Palomo, F., A., Macias, A., Blanco, M.T., Puertas, "Physical, chemical and mechanical characterization of geopolymers," in *Proceedings of the 9th International Congress on the Chemistry of Cement*, 1992, p. 505511.

[34] Rahier, H., B. Van Mele, M. Biesemans, J. Wastiels, and X. Wu, "Low-temperature synthesized aluminosilicate glasses: Part I. Low-temperature reaction stoichiometry and structure of a model compound," *J. Mater. Sci.*, vol. 31, no. 1, pp. 71–79, 1996, DOI: 10.1007/BF00355128.

[35] Van Jaarsveld, J. G. S., J. S. J. Van Deventer, and L. Lorenzen, "Factors affecting the immobilization of metals in polymerized flyash," *Metall. Mater. Trans. B Process Metall. Mater. Process. Sci.*, vol. 29, no. 1, pp. 283–291, 1998, DOI: 10.1007/s11663-998-0032-z.

[36] Hardjito, D., S. E. Wallah, D. M. J. Sumajouw, and B. V. Rangan, "On the development of fly ash-based geopolymer concrete," *ACI Mater. J.*, vol. 101, no. 6, pp. 467–472, 2004, DOI: 10.14359/13485.

[37] Moradikhou, A. B., A. Esparham, and M. Jamshidi Avanaki, "Physical & mechanical properties of fiber-reinforced metakaolin-based geopolymer concrete," *Constr. Build. Mater.*, vol. 251, p. 118965, 2020, DOI: 10.1016/j.conbuildmat.2020.118965.

[38] Rashad, A. M. "A comprehensive overview about the influence of different additives on the properties of alkali-activated slag - A guide for Civil Engineer," *Constr. Build. Mater.*, vol. 47, pp. 29–55, Oct. 2013, DOI: 10.1016/j.conbuildmat.2013.04.011.

[39] Davidovits, J. "Chemistry of geopolymeric systems, terminology," *Geopolymer*, vol. 99, no. 292, pp. 9–39, 1999.

[40] Zhang, M., N. Aaron Deskins, G. Zhang, R. T. Cygan, and M. Tao, "Modeling the Polymerization Process for Geopolymer Synthesis through Reactive Molecular Dynamics Simulations," *J. Phys. Chem. C*, vol. 122, no. 12, pp. 6760–6773, Mar. 2018, DOI: 10.1021/acs.jack.8b00697.

[41] Xu H. and J. S. J. Van Deventer, "The geopolymerisation of alumino-silicate minerals," *Int. J. Miner. Process.*, vol. 59, no. 3, pp. 247–266, Jun. 2000, DOI: 10.1016/S0301-7516(99)00074-5.

[42] Dixson, P., J. L. Provis, G. C. Lukey, and J. S. J. van Deventer, "The role of inorganic polymer technology in the development of 'green concrete,'" *Cem. Concr. Res.*, vol. 37, no. 12, pp. 1590–1597, Dec. 2007, DOI: 10.1016/j.cemconres.2007.08.018.

[43] Tho-In, T., V. Sata, P. Chindaprasirt, and C. Jaturapitakkul, "Previous high-calcium fly ash geopolymer concrete," *Constr. Build. Mater.*, vol. 30, pp. 366–371, May 2012, DOI: 10.1016/j.conbuildmat.2011.12.028.

[44] Matsagar, V. "Advances in structural engineering: Materials, volume three," *Adv. Struct. Eng. Mater. Vol. Three*, no. April, pp. 1619–2647, 2015, DOI: 10.1007/978-81-322-2187-6.

[45] Van Lam, T., B. Bulgakov, O. Aleksandrova, O. Larsen, and P. Ngoc Anh, "Effect of rice husk ash and fly ash on the compressive strength of high-performance concrete," *E3S Web Conf.*, vol. 33, pp. 1–13, 2018, DOI: 10.1051/e3sconf/20183302030.

[46] Duxson, P., A. A. Ferná Ndez-Jimé, A. J. L. Provis, A. G. C. Lukey, A. A. Palomo, and A. J. S. J. Van Deventer, "*Geopolymer technology: the current state of the art*," DOI: 10.1007/s10853-006-0637-z.

[47] Engin, Y. "*Çimento Üretiminde Kullanılan Mineral Katkılar*[*Mineral Additives Used in Cement Production*]," 2014. https://www.betonvecimento.com/surdurulebilirlik/mineral-katkilar (accessed Mar. 19, 2021).

[48] Kong D. L. Y. and J. G. Sanjayan, "Damage behavior of geopolymer composites exposed to elevated temperatures," *Cem. Concr. Compos.*, vol. 30, no. 10, pp. 986–991, Nov. 2008, doi: 10.1016/j.cemconcomp.2008.08.001.

[49] Sabir, B., S. Wild, and J. Bai, "Metakaolin and calcined clays as pozzolans for concrete: A review," *Cem. Concr. Compos.*, vol. 23, no. 6, pp. 441–454, Dec. 2001, doi: 10.1016/S0958-9465(00)00092-5.

[50] Rovnaník, P. "Effect of curing temperature on the development of the hard structure of metakaolin-based geopolymer," *Constr. Build. Mater.*, vol. 24, no. 7, pp. 1176–1183, Jul. 2010, DOI: 10.1016/j.conbuildmat.2009.12.023.

[51] Siddique R. and J. Klaus, "Influence of metakaolin on the properties of mortar and concrete: A review," *Applied Clay Science*, vol. 43, no. 3–4. pp. 392–400, Mar. 2009, DOI: 10.1016/j.clay.2008.11.007.

[52] Kim, H. S., S. H. Lee, and H. Y. Moon, "Strength properties and durability aspects of high strength concrete using Korean metakaolin," *Constr. Build. Mater.*, vol. 21, no. 6, pp. 1229–1237, Jun. 2007, DOI: 10.1016/j.conbuildmat.2006.05.007.

[53] Zhang M. H. and V. M. Malhotra, "Characteristics of a thermally activated alumino-silicate pozzolanic material and its use in concrete," *Cem. Concr. Res.*, vol. 25, no. 8, pp. 1713–1725, Dec. 1995, DOI: 10.1016/0008-8846(95)00167-0.

[54] Khatib J. M. and S. Wild, "Sulphate resistance of metakaolin mortar," *Cem. Concr. Res.*, vol. 28, no. 1, pp. 83–92, Jan. 1998, DOI: 10.1016/S0008-8846(97)00210-X.

[55] Khatib J. M. and R. M. Clay, "Absorption characteristics of metakaolin concrete," *Cem. Concr. Res.*, vol. 34, no. 1, pp. 19–29, Jan. 2004, DOI: 10.1016/S0008-8846(03)00188-1.

[56] Sawant, R. M. "*A Review on Experimental Investigation of Fibre Reinforced Metakaolin-Based Geopolymer Concrete*," no. July 2019.

[57] Rashad, A. M. "Metakaolin as cementitious material: History, scours, production and composition-A comprehensive overview," *Construction and Building Materials*, vol. 41. Elsevier, pp. 303–318, Apr. 01, 2013, DOI: 10.1016/j.conbuildmat.2012.12.001.

[58] Görür, E. "*Alkali Aktifleştirilmiş Uçucu Kül Geopolimer Betonun Dayanım Ve Durabilite Özelliklerinin Araştırılması,[Investigation of Strength and Durability Properties of Alkaline Activated Fly Ash Geopolymer Concrete*]" 2015.

[59] Ahmaruzzaman, M. "A review on the utilization of fly ash," *Progress in Energy and Combustion Science*, vol. 36, no. 3. Pergamon, pp. 327–363, Jun. 2010, DOI: 10.1016/j.pecs.2009.11.003.

[60] Joshi R. C. and R. P. Lohtia, "98/00323 Fly ash in concrete: Production, properties, and uses," *Fuel Energy Abstr.*, vol. 39, no. 1, p. 26, 1998, DOI: 10.1016/0140-6701(98)94358-2.

[61] Fernandez-Jimenez, A., I. García-Lodeiro, and A. Palomo, "Durability of alkali-activated fly ash cementitious materials," *J. Mater. Sci.*, vol. 42, no. 9, pp. 3055–3065, May 2007, DOI: 10.1007/s10853-006-0584-8.

[62] Wallah S. E. and B. V. Rangan, "Low-Calcium Fly Ash-Based Geopolymer Concrete: Long-Term Properties," pp. 1–107, 2006, [Online]. Available: https://espace.curtin.edu.au/handle/20.500.11937/34322.

[63] Ramezanianpour, A. A., M. A. Moeini, M. Bagheri, and H. Bayat, "An overview on the properties of Geopolymer concrete fabricated with industrial by-products," no. November 2016.

[64] Songpiriyakij, S., T. Kubprasit, C. Jaturapitakkul, and P. Chindaprasirt, "Compressive strength and degree of reaction of biomass- and fly ash-based geopolymer," *Constr. Build. Mater.*, vol. 24, no. 3, pp. 236–240, Mar. 2010, DOI: 10.1016/j.conbuildmat.2009.09.002.

[65] Fernández-Jiménez A. and A. Palomo, "Characterisation of fly ashes. Potential reactivity as alkaline cements," *Fuel*, vol. 82, no. 18, pp. 2259–2265, 2003, doi: 10.1016/S0016-2361(03)00194-7.

[66] Ryan M. and K. Ramyar, "*Temperature Resistance of Fly Ash-Based Geopolymer*," no. April 2017.

[67] Yousuf, A., S. O. Manzoor, M. Youssouf, Z. A. Malik, and K. Sajjad Khawaja, "*Fly Ash: Production and Utilization in India-An Overview*," 2020.

[68] Singh, N. B. "Fly ash-based geopolymer binder: A future construction material," *Minerals*, vol. 8, no. 7, 2018, DOI: 10.3390/min8070299.

[69] Part, W. K., M. Ramli, and C. B. Cheah, "An overview on the influence of various factors on the properties of geopolymer concrete derived from industrial by-products," *Construction and Building Materials*, vol. 77. Elsevier Ltd, pp. 370–395, Feb. 15, 2015, doi: 10.1016/j.conbuildmat.2014.12.065.

[70] Liew, Y. M., C. Y. Heah, A. B. Mohd Mustafa, and H. Kamarudin, "Structure and properties of clay-based geopolymer cement: A review," *Progress in Materials Science*, vol. 83. Elsevier Ltd, pp. 595–629, Oct. 01, 2016, DOI: 10.1016/j.pmatsci.2016.08.002.

[71] Caijun Shi, P. K., Della Roy, "*Alkali-Activated Cements and Concretes*," Taylor&Francis Group, Newyork, 2006. ISBNI3 :978-0-415-70004-7 (accessed Mar. 19, 2021).

[72] Siddique, R. "Utilization (recycling) of iron and steel industry by-product (GGBS) in concrete: Strength and durability properties," *J. Mater. Cycles Waste Manag.*, vol. 16, no. 3, pp. 460–467, 2014, DOI: 10.1007/s10163-013-0206-x.

[73] Mishra, J., S. K. Das, S. K. Singh, and S. M. Mustakim, "Development of Geopolymer Concrete for the Protection of Environment: A Greener Alternative to Cement," *Int. J. Civ. Eng.*, vol. 6, no. 3, pp. 41–47, 2019, DOI: 10.14445/23488352/ijce-v6i3p106.

[74] Saludung, A., Y. Ogawa, and K. Kawai, "Microstructure and mechanical properties of FA/GGBS-based geopolymer," in *MATEC Web of Conferences*, 2018, vol. 195, DOI: 10.1051/matecconf/201819501013.

[75] Ash-based geopolymer paste by incorporating ground granulated blast furnace slag," *Constr. Build. Mater.*, vol. 146, pp. 615–620, Aug. 2017, DOI: 10.1016/j.conbuildmat.2017.04.139.

[76] Al-Rawi S. and N. Tayşi, "Performance of self-compacting geopolymer concrete with and without GGBFS and steel fiber," *Adv. Concr. Constr.*, vol. 6, no. 4, pp. 323–344, Aug. 2018, DOI: 10.12989/acc.2018.6.4.323.

[77] Abhilash, P., C. Sashidhar, and I. V. Ramana Reddy, "Strength properties of fly ash and GGBS based geopolymer concrete," *Int. J. ChemTech Res.*, vol. 9, no. 3, pp. 350–356, 2016.

[78] Nagalingam G. and R. B. Chokkalingam, "Strength studies on geopolymer concrete with GGBS and Fly ash," *IOP Conf. Ser. Mater. Sci. Eng.*, vol. 872, no. 1, 2020, DOI: 10.1088/1757-899X/872/1/012107.

[79] Vinothkumar, A., M. Kalaivani, and P. Easwaran, "*Development of Fly ash-GGBS based Self Compacting Geopolymer Concrete : a Review*," no. September, pp. 1–5, 2019.

[80] Prabu, B., R. Kumutha, and K. Vijai, "Effect of fibers on the mechanical properties of fly ash and GGBS based geopolymer concrete under different curing conditions," *Indian J. Eng. Mater. Sci.*, vol. 24, no. 1, pp. 5–12, 2017.

[81] Kumar, D. N., K. Ramujee, and M. Tech, "Durability Characteristics of Fiber Reinforced Geopolymer Concrete Incorporated with Fly-ash and GGBS," *Int. Res. J. Eng. Technol.*, pp. 958–969, 2017.

[82] Yeğinobalı, A. *SİLİS DUMANI ve ÇİMENTO ile BETONDA KULLANIMI.* Ankara: TÜRKİYE ÇİMENTO MÜSTAHSİLLERİ BİRLİĞİ, 2009.[*USE IN CONCRETE WITH SILICONE SMOKE and CEMENT.* Ankara: CEMENT MANUFACTURERS ASSOCIATION OF TURKEY, 2009.]

[83] Rao, G. A. "Investigations on the performance of silica fume-incorporated cement pastes and mortars," *Cem. Concr. Res.*, vol. 33, no. 11, pp. 1765–1770, Nov. 2003, DOI: 10.1016/S0008-8846(03)00171-6.

[84] Khater, H. M. "Effect of silica fume on the characterization of the geopolymer materials," *Int. J. Adv. Struct. Eng.*, vol. 5, no. 1, pp. 1–10, 2013, DOI: 10.1186/2008-6695-5-12.

[85] Şimşek, O., A. Dur, H. Yaprak, G. Üniversitesi, T. E. Fakültesi, and Y. E. Bölümü, *"Silis dumanı ve süperakışkanlaştırıcı katkılı harçların özellikleri,*[Properties of mortars with silica fume and superplasticizer additives]" 2004.

[86] Okoye, F. N., J. Durgaprasad, and N. B. Singh, "Effect of silica fume on the mechanical properties of fly ash-based geopolymer concrete," *Ceram. Int.*, vol. 42, no. 2, pp. 3000–3006, 2016, DOI: 10.1016/j.ceramint.2015.10.084.

[87] Verma M. and N. Dev, "Sodium hydroxide effect on the mechanical properties of fly ash-slag based geopolymer concrete," *Struct. Concr.*, vol. 22, no. S1, pp. E368–E379, Jan. 2021, DOI: 10.1002/suco.202000068.

[88] Sravanthi, D., Y. Himath Kumar, and B. Sarath Chandra Kumar, "Comparative study on flow characteristics, strength, and durability of GGBS based geopolymer concrete," *IOP Conf. Ser. Mater. Sci. Eng.*, vol. 912, no. 6, 2020, DOI: 10.1088/1757-899X/912/6/062032.

[89] Ghazy, M. "SETTING AND MECHANICAL PROPERTIES OF FLY ASH BASED GEOPOLYMERS," in *International Conference on Advances in Structural and Geotechnical Engineering*, 2017, pp. 1–11.

[90] Komljenović, M., Z. Baščarević, and V. Bradić, "Mechanical and microstructural properties of alkali-activated fly ash geopolymers," *J. Hazard. Mater.*, vol. 181, no. 1–3, pp. 35–42, Sep. 2010, DOI: 10.1016/j.jhazmat.2010.04.064.

[91] Singh, B., G. Ishwarya, M. Gupta, and S. K. Bhattacharyya, "Geopolymer concrete: A review of some recent developments," *Construction and Building Materials*, vol. 85. Elsevier Ltd, pp. 78–90, Jun. 15, 2015, DOI: 10.1016/j.conbuildmat.2015.03.036.

[92] Esaifan, M., H. Khoury, I. Aldabsheh, H. Rahier, M. Hourani, and J. Wastiels, "Hydrated lime/potassium carbonate as an alkaline activating mixture to produce the kaolinitic clay-based inorganic polymer," *Appl. Clay Sci.*, vol. 126, pp. 278–286, Jun. 2016, DOI: 10.1016/j.clay.2016.03.026.

[93] Salih, M. A., A. A. Abang Ali, and N. Farzadnia, "Characterization of mechanical and microstructural properties of palm oil fuel ash geopolymer cement paste," *Constr. Build. Mater.*, vol. 65, pp. 592–603, Aug. 2014, DOI: 10.1016/j.conbuildmat.2014.05.031.

[94] Singh N. B. and B. Middendorf, "Geopolymers as an alternative to Portland cement: An overview," *Constr. Build. Mater.*, vol. 237, p. 117455, 2020, DOI: 10.1016/j.conbuildmat.2019.117455.

[95] Puertas, F., S. Martínez-Ramírez, S. Alonso, and T. Vázquez, "Alkali-activated fly ash/slag cements. Strength behaviour and hydration products," *Cem. Concr. Res.*, vol. 30, no. 10, pp. 1625–1632, Oct. 2000, doi: 10.1016/S0008-8846(00)00298-2.

[96] Hard todjito, D., C. C. Cheak, and C. H. Lee Ing, "Strength and Setting Times of Low Calcium Fly Ash-based Geopolymer Mortar," *Mod. Appl. Sci.*, vol. 2, no. 4, 2008, DOI: 10.5539/mas.v2n4p3.

[97] Yılmazer Polat, B., B. Yilmazer Polat, M. Uysal, and V. Korkmaz, "*A RESEARCH FOR BACTERIAL SELF-HEALING IN METAKAOLIN BASED GEOPOLYMER MORTARS*," 2020.

[98] Memon, F. A., M. F. Nuruddin, S. Khan, N. Shafiq, and T. Ayub, "Effect of sodium hydroxide concentration on fresh properties and compressive strength of self-compacting geopolymer concrete," *J. Eng. Sci. Technol.*, vol. 8, no. 1, pp. 44–56, 2013.

[99] Wang, H., H. Li, and F. Yan, "Synthesis and mechanical properties of metakaolinite-based geopolymer," *Colloids Surfaces A Physicochem. Eng. Asp.*, vol. 268, no. 1–3, pp. 1–6, Oct. 2005, DOI: 10.1016/j.colsurfa.2005.01.016.

[100] Kumaravel S. and K. Girija, "Acid and salt resistance of geopolymer concrete with varying concentration NaOH," *J. Eng. Res. Stud.*, no. May, pp. 3–5, 2013.

[101] Ersoy, U. "*Betonarme: temel ilkeler ve taşıma gücü hesabı,*[*Reinforced concrete: basic principles and bearing capacity calculation*]" 1985.

[102] Baradan, H., Bülent Serdar, Aydın Yazıcı, *Beton.*[Concrete] İzmir, 2012.

[103] Mishra J. and R. Panigrahi, "Mini-Review on Structural Performance of Fiber Reinforced Geopolymer Concrete," vol. 3, no. 2, pp. 435–442, 2020.

[104] Kocabeyoğlu, E. T. "*Bazalt Lif İçeren Yüksek Dayımlı Harçların Yüksek Sıcaklık Dirençleri,*[*High Temperature Resistance of High Strength Mortars Containing Basalt Fibers*]" Bozok Üniversitesi, 2016.

[105] ÖĞDÜ, M. K. "*Farklı Uzunluklarda Karbon Fiber Takviyeli Gazbetonun Mekanik ve Fiziksel Özelliklerinin Deneysel Olarak Araştırılması,*[*Experimental Investigation of Mechanical and Physical Properties of Carbon Fiber Reinforced Aerated Concrete of Different Lengths*]" 2018.

[106] "https://tr.wikipedia.org/wiki/Kevlar".

[107] Ranjbar, N., S. Talebian, M. Mehrauli, C. Kuenzel, H. S. Cornelis Metselaar, and M. Z. Jumaat, "Mechanisms of interfacial bond in steel and polypropylene fiber reinforced geopolymer composites," *Compos. Sci. Technol.*, vol. 122, pp. 73–81, 2016, DOI: 10.1016/j.compscitech.2015.11.009.

[108] Ranjbar N. and M. Zhang, "Fiber-reinforced geopolymer composites: A review," *Cem. Concr. Compos.*, vol. 107, no. November 2019, p. 103498, 2020, doi: 10.1016/j.cemconcomp.2019.103498.

[109] Nataraja, M. C., N. Dhang, and A. P. Gupta, "Toughness characterization of steel fiber-reinforced concrete by JSCE approach," *Cem. Concr. Res.*, vol. 30, no. 4, pp. 593–597, 2000, doi: 10.1016/S0008-8846(00)00212-X.

[110] Çivici, F. "Çelik Lif Donatılı Betonun Eğilme Tokluğu," *Pamukkale Üniversitesi Mühendislik Bilim. Derg.*,[Flexural Toughness of Steel Fiber Reinforced Concrete," *Pamukkale University Engineering Science. Journal*] vol. 12, no. 2, pp. 183–188, 2006, doi: 10.11.2004.

[111] Branston, J., S. Das, S. Y. Kenno, and C. Taylor, "Mechanical behavior of basalt fiber reinforced concrete," *Constr. Build. Mater.*, vol. 124, pp. 878–886, Oct. 2016, DOI: 10.1016/j.conbuildmat.2016.08.009.

[112] Song, P. S., S. Hwang, and B. C. Sheu, "Strength properties of nylon- and polypropylene-fiber-reinforced concrete," *Cem. Concr. Res.*, vol. 35, no. 8, pp. 1546–1550, Aug. 2005, doi: 10.1016/j.cemconres.2004.06.033.

[113] Nili M. and V. Afroughsabet, "The effects of silica fume and polypropylene fibers on the impact resistance and mechanical properties of concrete," *Constr. Build. Mater.*, vol. 24, no. 6, pp. 927–933, Jun. 2010, DOI: 10.1016/j.conbuildmat. 2009.11.025.

[114] Junaid, M. T., A. Elbana, and S. Altoubat, "Flexural response of geopolymer and fiber-reinforced geopolymer concrete beams reinforced with GFRP bars and strengthened using CFRP sheets," *Structures*, vol. 24, no. February, pp. 666–677, 2020, DOI: 10.1016/j.istruc.2020.02.003.

[115] Wang, Z. L., Z. M. Shi, and J. G. Wang, "On the strength and toughness properties of SFRC under static-dynamic compression," *Compos. Part B Eng.*, vol. 42, no. 5, pp. 1285–1290, Jul. 2011, DOI: 10.1016/j.compositesb.2011.01.027.

[116] Millard, S. G., T. C. K. Molyneaux, S. J. Barnett, and X. Gao, "Dynamic enhancement of blast-resistant ultra-high-performance fiber-reinforced concrete under flexural and shear loading," *Int. J. Impact Eng.*, vol. 37, no. 4, pp. 405–413, Apr. 2010, DOI: 10.1016/j.ijimpeng.2009.09.004.

[117] Yap, S. P., U. J. Alengaram, and M. Z. Jumaat, "Enhancement of mechanical properties in polypropylene- and nylon-fiber reinforced oil palm shell concrete," *Mater. Des.*, vol. 49, pp. 1034–1041, Aug. 2013, DOI: 10.1016/ j.matdes.2013.02.070.

[118] Zhang P. and Q. F. Li, "Effect of polypropylene fiber on the durability of the concrete composite containing fly ash and silica fume," *Compos. Part B Eng.*, vol. 45, no. 1, pp. 1587–1594, Feb. 2013, doi: 10.1016/j.compositesb. 2012.10.006.

[119] Ma, Y., B. Zhu, and M. Tan, "Properties of ceramic fiber reinforced cement composites," DOI: 10.1016/j.cemconres.2004.05.017.

[120] Bhalchandra S. A. and A. Y. Bhosle, *"Properties of Glass Fibre Reinforced Geopolymer Concrete,"* vol. 3, pp. 2007–2010, 2013.

[121] kürşat esat Alyamac, *"Toz Katkı Tipinin Kendiliğinden Yerleşen Betonun Kırılganlığa Etkisi*[*Effect of Powder Additive Type on Fragility of Self-Compacting Concrete*]," 2011.

[122] Chu, L., S H/Ye, H/Huang, L/Li, "*Carbon fiber reinforced geopolymer (FRG) mix design based on liquid film thickness*," 2020.

[123] Samal, S., N. Phan Thanh, I. Petríková, B. Marvalová, K. A. M. Vallons, and S. V. Lomov, "Correlation of microstructure and mechanical properties of various fabric reinforced geo-polymer composites after exposure to elevated temperature," *Ceram. Int.*, vol. 41, no. 9, pp. 12115–12129, 2015, DOI: 10.1016/ j.ceramint.2015.06.029.

[124] Wen S. and D. D. L. Chung, "Electric polarization in carbon fiber-reinforced cement," *Cem. Concr. Res.*, vol. 31, no. 1, pp. 141–147, Jan. 2001, doi: 10.1016/S0008-8846(00)00382-3.

[125] Salih yazıcıoğlu bahar demirel, *"Silis Dumanının Karbon Fiber Takviyeli Hafif Betonun Mekanik Özelliklerine Etkisi* [*The effect of silica fume on the mechanical properties of carbon fiber reinforced lightweight concrete*,]" vol. 1, pp. 103–109, 2007.

[126] Di Ludovico, G. M., Marco, Andrea Prota, "Structural upgrade using basalt fibers for concrete confinement," *J. Compos. Constr.*, vol. 14, no. 5, pp. 541–552, 2010.

[127] Kabay, N. "Abrasion resistance and fracture energy of concretes with basalt fiber," *Construction and Building Materials*, vol. 50. pp. 95–101, 2014, DOI: 10.1016/j.conbuildmat.2013.09.040.

[128] Sim, J., C. Park, and D. Y. Moon, "Characteristics of basalt fiber as a strengthening material for concrete structures," *Compos. Part B Eng.*, vol. 36, no. 6–7, pp. 504–512, 2005, DOI: 10.1016/j.compositesb.2005.02.002.

[129] Militký, J., V. Kovačič, and J. Rubnerová, "Influence of thermal treatment on the tensile failure of basalt fibers," *Eng. Fract. Mech.*, vol. 69, no. 9, pp. 1025–1033, 2002, DOI: 10.1016/S0013-7944(01)00119-9.

[130] Kızılkanat, G., AB; Kabay, N; Akyüncü, V; Erdoğan, "Basalt Fibers and Mechanical Properties of Basalt Fiber," *Sigma J. Eng. Nat. Sci. ve Fen Bilim. Derg.*, vol. 4, no. 32, pp. 444–452, 2014.

[131] Li W. and J. Xu, "Impact Characterization of basalt fiber reinforced geopolymeric concrete using a 100-mm-diameter split Hopkinson pressure bar," *Mater. Sci. Eng. A* vol. 513–514, no. C, pp. 145–153, Jul. 2009, DOI: 10.1016/j.msea. 2009.02.033.

[132] Matthew F. and R. Rawlings, "Composite Materials: Engineering and Science - F L Matthews, R D Rawlings - Google Books," *Woodhead Publishing*, 1999. (accessed Mar. 30, 2021), ISBN : 184569855X, 9781845698553.

[133] Agarwal, B. D., L. J. Broutman, and C. W. Bert, "Analysis and Performance of Fiber Composites," *J. Appl. Mech.*, vol. 48, no. 1, pp. 213–213, 1981, DOI: 10.1115/1.3157582.

[134] In: A. R. B. (Eds. . H. H. Yang, "Fiber Reinforcement of Composite Materials," in *Elsevier*, 1988, pp. 246–329.

[135] Bunsell, A. R. *Handbook of Tensile Properties of Textile and Technical Fibres*. 2009.

[136] Young, R. J., D. Lu, R. J. Day, W. F. Knoff, and H. A. Davis, "Relationship between structure and mechanical properties for aramid fibers," *J. Mater. Sci.*, vol. 27, no. 20, pp. 5431–5440, 1992, DOI: 10.1007/BF00541602.

[137] "Kevlar nedir? Nasıl üretilir? Özellikleri nedir?" https://malzemebilimi. net/kevlar-nedir-nasil-retailer-ozellikleri-nedir.html (accessed Apr. 01, 2021).

[138] Cheng, M., W. Chen, and T. Weerasooriya, "Mechanical properties of Kevlar® KM2 single fiber," *J. Eng. Mater. Technol. Trans. ASME*, vol. 127, no. 2, pp. 197–203, 2005, DOI: 10.1115/1.1857937.

[139] Bencomo-Cisneros J. A. et al., "Characterization of Kevlar-29 fibers by tensile tests and nanoindentation," *J. Alloys Compd.*, vol. 536, no. SUPPL.1, pp. S456–S459, 2012, DOI: 10.1016/j.jallcom.2011.11.031.

[140] Deteresa, R. S. P. S. J., S. R. Allen, R. J. Farris, "Compressive and torsional behavior of Kevlar 49 fiber," *J. Mater. Sci.*, vol. 19, pp. 57–72, 1984.

[141] Zhu, D., B. Mobasher, A. Vaidya, and S. D. Rajan, "Mechanical behaviors of Kevlar 49 fabric subjected to uniaxial, biaxial tension and in-plane large shear deformation," *Compos. Sci. Technol.*, vol. 74, pp. 121–130, Jan. 2013, doi: 10.1016/j.compscitech.2012.10.012.

[142] Yoo, D. Y., J. H. Lee, and Y. S. Yoon, "Effect of fiber content on mechanical and fracture properties of ultra-high-performance fiber-reinforced cementitious composites," *Compos. Struct.*, vol. 106, pp. 742–753, Dec. 2013, DOI: 10.1016/J.COMPSTRUCT.2013.07.033.

[143] Karunanithi S. and S. Anandan, "*Flexural Toughness Properties of Reinforced Steel Fibre Incorporated Alkali Activated Slag Concrete*," 2014, DOI: 10.1155/2014/719436.

[144] Bhutta, A., P. H. R. Borges, C. Zanotti, M. Farooq, and N. Banthia, "Flexural behavior of geopolymer composites reinforced with steel and polypropylene macro fibers," *Cem. Concr. Compos.*, vol. 80, pp. 31–40, Jul. 2017, DOI: 10.1016/J.CEMCONCOMP.2016.11.014.
[145] Liu, Y., Z. Zhang, C. Shi, D. Zhu, N. Li, and Y. Deng, "Development of ultra-high performance geopolymer concrete (UHPGC): Influence of steel fiber on mechanical properties," *Cem. Concr. Compos.*, vol. 112, p. 103670, Sep. 2020, DOI: 10.1016/j.cemconcomp.2020.103670.
[146] Yoo, D. Y., S. T. Kang, and Y. S. Yoon, "Effect of fiber length and placement method on flexural behavior, tension-softening curve, and fiber distribution characteristics of UHPFRC," *Constr. Build. Mater.*, vol. 64, pp. 67–81, Aug. 2014, DOI: 10.1016/J.CONBUILDMAT.2014.04.007.
[147] Aydin S. and B. Baradan, "The effect of fiber properties on high-performance alkali-activated slag/silica fume mortars," *Compos. Part B Eng.*, vol. 45, no. 1, pp. 63–69, Feb. 2013, DOI: 10.1016/J.COMPOSITESB.2012.09.080.
[148] Şimşek, O. "Lifli Beton[Fiber Concrete]." https://insapedia.com/lifli-beton/.
[149] Sezik, M. "Kevlar." https://muratsezik.wordpress.com/2015/12/05/kevlar/ (accessed Apr. 01, 2021).
[150] Wikipedia, "Basalt Fiber." https://en.wikipedia.org/wiki/Basalt_fiber.
[151] Ranjbar, N., M. Mehrali, M. R. Maheri, and M. Mehrali, "Hot-pressed geopolymer," *Cem. Concr. Res.*, vol. 100, pp. 14–22, Oct. 2017, DOI: 10.1016/j.cemconres.2017.05.010.
[152] Okoye, F. N., J. Durgaprasad, and N. B. Singh, "Mechanical properties of alkali-activated fly ash/Kaolin based geopolymer concrete," *Constr. Build. Mater.*, vol. 98, pp. 685–691, Nov. 2015, DOI: 10.1016/j.conbuildmat.2015.08.009.
[153] Yuan J. et al., "Effect of curing temperature and SiO2/K2O molar ratio on the performance of metakaolin-based geopolymers," *Ceram. Int.*, vol. 42, no. 14, pp. 16184–16190, Nov. 2016, doi: 10.1016/j.ceramint.2016.07.139.
[154] Matli, R. R. R. Panchal Reddy, Adel Mohamed Amer Mohamed, *Microwave Fast Sintering of Double Perovskite Ceramic Materials*. Advanced Ceramic Processing, 2015.
[155] Fareed Ahmed, M., M. Fadhil Nuruddin, and N. Shafiq, "Compressive strength and workability characteristics of low-calcium fly ash-based self-compacting geopolymer concrete," *World Acad. Sci. Eng. Technol.*, vol. 74, no. February, pp. 8–14, 2011.
[156] Nuruddin, M. F., S. Demie, and N. Shafiq, "Effect of mix composition on workability and compressive strength of self-compacting geopolymer concrete," *Can. J. Civ. Eng.*, vol. 38, no. 11, pp. 1196–1203, Nov. 2011, DOI: 10.1139/l11-077.
[157] Verma M. and N. Dev, "Effect of ground granulated blast furnace slag and fly ash ratio and the curing conditions on the mechanical properties of geopolymer concrete," *Struct. Concr.*, no. August 2020, pp. 1–15, 2021, DOI: 10.1002/suco.202000536.
[158] Venkatesan R. P. and K. C. Pazhani, "Strength and durability properties of geopolymer concrete made with Ground Granulated Blast Furnace Slag and Black

Rice Husk Ash," *KSCE J. Civ. Eng.*, vol. 20, no. 6, pp. 2384–2391, Sep. 2016, DOI: 10.1007/s12205-015-0564-0.

[159] Patel Y. J. and N. Shah, "Development of self-compacting geopolymer concrete as a sustainable construction material," *Sustain. Environ. Res.*, vol. 28, no. 6, pp. 412–421, Nov. 2018, DOI: 10.1016/j.serj.2018.08.004.

[160] Al-Shether, B., T. Al-Attar, and Z. A. Hassan, "Effect of Curing System on Metakaolin Based Geopolymer Concrete," *J. Univ. Babylon - Eng. Sci.*, vol. 24, no. 3, pp. 569–576, 2016.

[161] Chandra Padmakar K. and B. Sarath Chandra Kumar, "An experimental study on metakaolin and GGBS based geopolymer concrete," *Int. J. Civ. Eng. Technol.*, vol. 8, no. 1, pp. 544–557, 2017, doi: 10.21817/ijet/2017/v9i1/170902311.

[162] Celik, A., K. Yilmaz, O. Canpolat, M. M. Al-mashhadani, Y. Aygörmez, and M. Uysal, "High-temperature behavior and mechanical characteristics of boron waste additive metakaolin based geopolymer composites reinforced with synthetic fibers," *Constr. Build. Mater.*, vol. 187, pp. 1190–1203, Oct. 2018, DOI: 10.1016/j.conbuildmat.2018.08.062.

[163] Aygörmez, Y., O. Canpolat, M. M. Al-mashhadani, and M. Uysal, "Elevated temperature, freezing-thawing and wetting-drying effects on polypropylene fiber reinforced metakaolin based geopolymer composites," *Construction and Building Materials*, vol. 235. 2020, DOI: 10.1016/j.conbuildmat.2019.117502.

[164] Keerthy V. and Y. H. Kumar, "Experimental studies on properties of geopolymer concrete with GGBS and fly ash," *Int. J. Civ. Eng. Technol.*, vol. 8, no. 1, pp. 602–609, 2017.

[165] R. ince. Erdinç Arıcı, Resul Dursun, "*Betonun çarpma mukavemetinin tesbiti*,[*Determination of the impact strength of concrete*]" 2007, no. November, pp. 628–633.

[166] Zhu, X. C., H. Zhu, and H. R. Li, "Drop-weight impact test on U-shape concrete specimens with statistical and regression analyses," *Materials (Basel).*, vol. 8, no. 9, pp. 5877–5890, 2015, DOI: 10.3390/ma8095281.

[167] Zhang, M. H., V. P. W. Shim, G. Lu, and C. W. Chew, "Resistance of high-strength concrete to projectile impact," *Int. J. Impact Eng.*, vol. 31, no. 7, pp. 825–841, Aug. 2005, DOI: 10.1016/j.ijimpeng.2004.04.009.

[168] Vossoughi, F., C. P. Ostertag, P. J. M. Monteiro, and G. C. Johnson, "Resistance of concrete protected by fabric to projectile impact," *Cem. Concr. Res.*, vol. 37, no. 1, pp. 96–106, Jan. 2007, doi: 10.1016/j.cemconres.2006.09.003.

[169] Yankelevsky, D. Z. "Local response of concrete slabs to low-velocity missile impact," *Int. J. Impact Eng.*, vol. 19, no. 4, pp. 331–343, Apr. 1997, DOI: 10.1016/S0734-743X(96)00041-3.

[170] Soufeiani, L., S. N. Raman, M. Z. Bin Jumaat, U. J. Alengaram, G. Ghadyani, and P. Mendis, "Influences of the volume fraction and shape of steel fibers on fiber-reinforced concrete subjected to dynamic loading – A review," *Engineering Structures*, vol. 124. Elsevier Ltd, pp. 405–417, Oct. 01, 2016, DOI: 10.1016/ j.engstruct.2016.06.029.

[171] Al-Tayeb, M. M., B. H. A. Bakar, H. Ismail, and H. M. Akil, "Impact Resistance of Concrete with Partial Replacements of Sand and Cement by Waste Rubber,"

Polym. - Plast. Technol. Eng., vol. 51, no. 12, pp. 1230–1236, 2012, DOI: 10.1080/03602559.2012.696767.

[172] Yunsheng, Z., S. Wei, and L. Zongjin, "Impact behavior and microstructural characteristics of PVA fiber reinforced fly ash-geopolymer boards prepared by extrusion technique," *Journal of Materials Science*, vol. 41, no. 10. pp. 2787–2794, 2006, DOI: 10.1007/s10853-006-6293-5.

[173] Dattatreya, M. C. K, J. K., and P. S. Ambily, "Investigation on the behavior of reinforced geopolymer concrete slab under repeated low-velocity impact loading," *Int. J. Innov. Res. Sci. Eng. Technol. (An ISO Certif. Organ.*, vol. 3, no. 3, pp. 10775–10786, 2014, Accessed: Nov. 25, 2020. [Online]. Available: www.ijirset.com.

[174] Nagan S. and R. Mohana, "Behaviour of geopolymer ferrocement slabs subjected to impact," *Iran. J. Sci. Technol. - Trans. Civ. Eng.*, vol. 38, no. C1+, pp. 223–233, 2014.

[175] Al-Hadithi, A. I., K. I. Aziz, and M. T. N. Al-Dulaim, "Behavior of Ferro-cement slabs modified by polymer under the low-velocity impact," in *Advanced Materials Research*, 2014, vol. 925, pp. 3–7, DOI: 10.4028/www.scientific.net/AMR.925.3.

[176] Dong Z. et al. "*Dynamic Behaviors of FA-GGBS-HMNS Based Geopolymer Paste When Subjected to Impact Compressive Loadings*," 2019.

[177] Abubakr, A. E., A. M. Soliman, and S. H. Diab, "Effect of activator nature on the impact behavior of Alkali-Activated slag mortar," *Constr. Build. Mater.*, vol. 257, p. 119531, 2020, DOI: 10.1016/j.conbuildmat.2020.119531.

[178] Duc Ngo, T. "Blast loading and blast effects on structures-an overview Advanced structural systems for prefabricated modular buildings *Hydrogen Embrittlement-Understanding and research framework*."

[179] Agrawal, J. P. *High Energy Materials: Propellants, Explosives, and Pyrotechnics*. 2010.

[180] Mortar, N. A. M., K. Hussin, R. A. Razak, M. M. A. B. Abdullah, A. H. Hilmi, and A. V. Sandu, "Properties and Behavior of Geopolymer Concrete Subjected to Explosive Air Blast Loading: A Review," in *MATEC Web of Conferences*, 2017, vol. 97, DOI: 10.1051/matecconf/20179701019.

[181] Shaikh F. U. A. and V. Vimonsatit, "Compressive strength of fly-ash-based geopolymer concrete at elevated temperatures," *Fire Mater.*, vol. 39, no. 2, pp. 174–188, Mar. 2015, DOI: 10.1002/fam.2240.

[182] Al-Bakri Abdullah, M. M., L. Jamaluddin, K. Hussin, M. Bnhussain, C. M. R. Ghazali, and M. I. Ahmad, "Fly ash porous material using geopolymerization process for high-temperature exposure," *Int. J. Mol. Sci.*, vol. 13, no. 4, pp. 4388–4395, Apr. 2012, DOI: 10.3390/ijms13044388.

[183] Yao, S. J., D. Zhang, F. Y. Lu, W. Wang, and X. G. Chen, "Damage features and dynamic response of RC beams under the blast," *Eng. Fail. Anal.*, vol. 62, pp. 103–111, Apr. 2016, DOI: 10.1016/j.engfailanal.2015.12.001.

[184] Meng, Q., C. Wu, Y. Su, J. Li, J. Liu, and J. Pang, "Experimental and numerical investigation of blast resistant capacity of high-performance geopolymer concrete panels," *Compos. Part B Eng.*, vol. 171, no. September 2018, pp. 9–19, 2019, DOI: 10.1016/j.compositesb.2019.04.010.

[185] Q. Meng et al. "Steel fiber reinforced alkali-activated geopolymer concrete slabs subjected to a natural gas explosion in buried utility tunnel," *Constr. Build. Mater.*, vol. 246, p. 118447, Jun. 2020, DOI: 10.1016/j.conbuildmat.2020.118447.

[186] Meng, Q., C. Wu, Y. Su, J. Li, J. Liu, and J. Pang, "A study of steel wire mesh reinforced high-performance geopolymer concrete slabs under blast loading," *J. Clean. Prod.*, vol. 210, pp. 1150–1163, Feb. 2019, DOI: 10.1016/j.jclepro. 2018.11.083.

[187] Cai H. and X. Liu, "Freeze-thaw durability of concrete: Ice formation process in pores," *Cem. Concr. Res.*, vol. 28, no. 9, pp. 1281–1287, Sep. 1998, DOI: 10.1016/S0008-8846(98)00103-3.

[188] Wang, Y., F. Gong, D. Zhang, and T. Ueda, "Estimation of ice formation in mortar saturated with sodium chloride solutions," *Constr. Build. Mater.*, vol. 144, pp. 238–251, Jul. 2017, DOI: 10.1016/j.conbuildmat.2017.03.073.

[189] Gong, F., E. Sicat, D. Zhang, and T. Ueda, "Stress analysis for concrete materials under multiple freeze-thaw cycles," *J. Adv. Concr. Technol.*, vol. 13, no. 3, pp. 124–134, Mar. 2015, DOI: 10.3151/jact.13.124.

[190] Zhao, J., G. Cai, D. Gao, and S. Zhao, "Influences of freeze-thaw cycle and curing time on chloride ion penetration resistance of Sulphoaluminate cement concrete," *Constr. Build. Mater.*, vol. 53, pp. 305–311, Feb. 2014, DOI: 10.1016/ j.conbuildmat.2013.11.110.

[191] geopolymer concrete," *Constr. Build. Mater.*, vol. 222, pp. 474–483, Oct. 2019, DOI: 10.1016/j.conbuildmat.2019.06.166.

[192] Yuan Y. et al., "Frost resistance of fiber-reinforced blended slag and Class F fly ash-based geopolymer concrete under the coupling effect of freeze-thaw cycling and axial compressive loading," *Constr. Build. Mater.*, vol. 250, p. 118831, Jul. 2020, DOI: 10.1016/j.conbuildmat.2020.118831.

[193] Mehta A. and R. Siddique, "Sulfuric acid resistance of fly ash-based geopolymer concrete," *Constr. Build. Mater.*, vol. 146, pp. 136–143, Aug. 2017, DOI: 10.1016/j.conbuildmat.2017.04.077.

[194] Ariffin, M. A. M., M. A. R. Bhutta, M. W. Hussin, M. Mohd Tahir, and N. Aziah, "Sulfuric acid resistance of blended ash geopolymer concrete," *Constr. Build. Mater.*, vol. 43, pp. 80–86, Jun. 2013, DOI: 10.1016/j.conbuildmat.2013.01.018.

[195] Bakharev, T. "Durability of geopolymer materials in sodium and magnesium sulfate solutions," *Cem. Concr. Res.*, vol. 35, no. 6, pp. 1233–1246, Jun. 2005, DOI: 10.1016/j.cemconres.2004.09.002.

[196] Ganesan, N., R. Abraham, and S. Deepa Raj, "Durability characteristics of steel fiber reinforced geopolymer concrete," *Constr. Build. Mater.*, vol. 93, pp. 471–476, 2015, DOI: 10.1016/j.conbuildmat.2015.06.014.

Chapter 4

Development and Application of Sulfated Geopolymer Catalyst in the Oxidation of Dibenzothiophene

Snehlata Kumari, Biswajit Saha and Sonali Sengupta*
Department of Chemical Engineering, Indian Institute of Technology Kharagpur, Kharagpur, West Bengal, India

Abstract

A sulfated geopolymer catalyst was prepared via a three-staged treatment comprising emulsion templating, ion exchange and sulfation treatment. The modification method improved the catalytic performance, surface morphology and porosity. Emulsion templating aided in creating macro- and mesoporosity in the sample, while ion exchange and sulfation treatments enhanced the acidity of the catalyst. The catalysts obtained at each stage of the treatment were characterized by SEM, TEM, XRD, XRF, BET, NH_3-TPD, and FTIR analyses. The catalysts prepared at each stage of the sequential treatment were subjected to a dibenzothiophene oxidation reaction, a model reaction for studying the applicability of the sulfated geopolymer catalyst in oxidative desulfurization. The efficiency of the catalyst was enhanced after each stage, and the conversion of dibenzothiophene was increased from 39 to 74%. The catalyst was thus deemed suitable for the preliminary treatment of petroleum fuel for reduction of sulfur content.

Keywords: dibenzothiophene, geopolymer, oxidation, catalysis

* Corresponding Author's Email: sonalis.@che.iitkgp.ac.in, sonalis.iitkgp@gmail.com.

In: Geopolymers
Editor: 979-8-88697-197-2
ISBN: Joseph E. Glenn

Introduction

Geopolymers, proposed by Joseph Davidovits in 1978, are inorganic and amorphous three-dimensional materials prepared by low-temperature reaction of alumino-silicates with alkali activators (Davidovits 1991). They consist of cross-linked SiO_4 and AlO^{4-} structural units connected to polymer structure through corner-sharing oxygen atoms (Brus et al. 2016). The net negative charge of the polymer structure is counterbalanced by charge balancing alkali cations, like K^+, Na^+ or Ca^{2+}. Geopolymers can be prepared from several alumino-silicate sources, viz., calcined clay, blast furnace slag, ground glass fiber, steel slag, fly ash, red clay ceramic powders, etc. (Kovářík et al. 2017) (Keppert et al. 2018) (Y. J. Zhang et al. 2017) (Wan, Rao, and Song 2017) (Y. Chen et al. 2019) (Dezfouli H R, Rangaraju P R 2018). Conventionally, geopolymers were mainly utilized as a substitute for Ordinary Portland Cement (OPC). OPC production involves the calcination of limestone, a key cause of approximately 7% of the worldwide CO_2 emissions (Metz et al. 2005). With the development of geopolymer cement, the CO_2 emission due to OPC production has been reported to decrease by almost 80%, making geopolymers the best-suited alternative to curb CO_2 emissions caused by the production of OPC (Akbari, Mensah-biney, and Simms 2015). Apart from that, geopolymers also act as eco-friendly materials that can be synthesized by various industrial and agricultural wastes such as fly ash, ground glass fiber, red mud, blast furnace slag, rice husk, etc., ultimately leading to a reduction in mining activities and the utilization of natural minerals.

Conventionally, geopolymers were majorly used in the construction industry, but lately, their usage has extended beyond the construction industry. They have found varied applications as adhesives, acoustic and thermal insulators, catalyst and membrane supports, adsorbents, catalysts, and for the solidification of hazardous wastes (Keppert et al. 2018). Although they have gathered immense attention from the research community due to their high thermal, mechanical and chemical resistance, they still require tremendous study to make them suitable for catalytic and adsorptive applications. The small pore size of geopolymeric materials hinders the diffusion of bulky molecules inside the pores, decreasing its efficiency as a catalyst or an adsorbent. The pore size of geopolymers can be tuned, and hierarchical porosity can be introduced into geopolymers, making the diffusion of bulky molecules and accessibility to active centers convenient. In recent times, varied methods of preparation of highly porous geopolymers having porosities as high as ≥ 50 vol% and bulk densities as lower as ≤ 0.7 g/cm^3 have been devised (Bai and Colombo 2018). Some of those techniques include solid templating using carbonaceous or biological materials, soft templating using various emulsifying agents, or non-templating methods based on the aggregation of nano-crystals or controlled crystallization of amorphous gels (Alzeer and MacKenzie 2018). Other methods for porosity development include demetallation (desilication or dealumination) or delamination. However, all the above-listed processes have their limitations, and they must be used accordingly. Geopolymers can also undergo ion exchange by exchanging monovalent alkali cations present in them to produce geopolymers with zeolite-like properties. Their compositional similitude with zeolites reflects their capability to be used as efficient catalysts with required modifications. The selection of raw material is crucial for preparing geopolymers with desired properties. Metakaolin-based geopolymers find wide applications in coatings, adhesives, and hydro-ceramics. Fly ash geopolymers possess high durability and mechanical strength. Geopolymers can also be modified by incorporating nanomaterials such as nano-clay, nanotubes, and nano-SiO_2. Still, as uniform dispersion of the nanomaterials throughout the geopolymer dictates the end properties of the synthesized material, it is both challenging and crucial (Asim et al. 2019). Homogeneous dispersion of nanomaterials in the geopolymer facilitates control over the pore size and surface area, which are important for catalyst performance. In addition to all these, geopolymers can also be transformed into acid/base catalysts which find varied industrial applications and are gaining interest due to their facile preparation methods, limited disposal issues, non-corrosiveness, and ability to be reused and regenerated (Asim et al. 2019).

Geopolymers have been extensively used for adsorptive applications and are now being used for catalytic applications. Some of the recent catalytic applications include transesterification, alkylation, benzylation, Beckmann rearrangement reaction, dibenzothiophene (DBT) oxidation reaction, dye degradation, reduction of NO_x and oxidation of VOCs.(Sharma et al. 2015) (Alzeer, MacKenzie, and Keyzers 2017) (Alzeer and MacKenzie 2018) (Alzeer, MacKenzie, and Keyzers 2016) (Kumari et al. 2021) (Y. Zhang and Liu 2013) (Sazama et al. 2011). In all the cases, the geopolymers were first made porous to facilitate better diffusion of reactants and were then modified as per the target application. This work involves the preparation of a sulfated geopolymer catalyst for a model oxidative desulfurization reaction studied via oxidation of DBT. The growing reduction of the share of conventional crudes out of the total oil resources remarks the grave need to develop feasible sulfur reduction processes. The reduction of sulfur content in crudes can be attained via hydrodesulfurization, oxidative desulfurization, adsorptive desulfurization, extractive desulfurization, bio desulfurization, and membrane desulfurization processes. While hydrodesulfurization is a conventional desulfurization technique requiring severe operating conditions, oxidative, adsorptive, extractive, and bio desulfurization are non-conventional techniques operating at mild conditions. The oxidative desulfurization technique relies on the fundamental difference in the polarity of the reactants and the products. At first, the existing sulfur compounds are oxidized to their corresponding sulfoxides/sulfones and are then withdrawn from the oil via adsorption or extraction with polar solvents. Oxidants such as tert-butyl hydrogen peroxide (TBHP), hydrogen peroxide (H_2O_2), ozone and molecular oxygen have been commonly employed in oxidative desulfurization reaction. As catalyst plays a vital role in the reaction, various catalysts have been developed so far. Metal impregnated catalysts such as Ti-Mo impregnated nano Hap, MoO_2 impregnated graphitic carbon nitride (gC_3N_4) nanocomposite, Fe(III) porphyrins, microporous titania-silica (TiO_2–SiO_2) nano-composites, $H_3PW_{12}O_{40}/TiO_2$ nanofiber catalyst, etc. have been employed in various desulfurization reactions (Saha, Yadav, and Sengupta 2018) (K. Chen et al. 2018) (Zhao et al. 2017) (Bazyari et al. 2016) (Ma et al. 2016). These catalysts are economically draining due to the use of costly metallic ingredients. So, developing cheap and efficient catalysts for the desulfurization of fuel oil is essential for obtaining significant revenues.

Solid superacid catalysts are emerging as an efficient replacement for HF-based liquid superacids that work efficiently at standard conditions of temperature but are not reasonably suited for usage in industries considering the difficulties in separation (Arata 1990). Also, HF-based liquid superacid catalysts are incompatible with environmental regulations (Arata 1990). Conventional acid catalysts used by industries such as $AlCl_3$, H_2SO_4, and BF_3 are highly corrosive. Thus, searching for an environmentally benign catalyst that can act as a suitable substitute is often sought after. Sulfated catalysts such as sulfated titania, sulfated zirconia, and sulfated iron oxide catalysts exhibit high acidity, thermal stability, and excellent catalytic activity and thus have attracted major attention from the researcher community (Arata 1990) (J. Sohn and Lee 2004). These catalysts can promote reactions such as cracking, alkylation and isomerization. Such catalysts are expected to work effectively to remove basic sulfur compounds from fuel oils. Geopolymer-based catalysts have been rarely utilized in desulfurization reactions. Therefore, this work focused on developing and evaluating the activity of a sulfated geopolymer catalyst prepared via three-staged treatment for a model oxidative desulfurization reaction.

Experimental Work

Materials

Kiran Global Green Cement, India supplied the geopolymer precursor (GP) and alkali activator (AA). Laboratory grade hexane was procured from Simson Pharma, India. Isooctane and analytical grade sulfuric acid (98%) were procured from Merck, India. Spectrochem Private Ltd, India provided tertiary butyl hydroperoxide (TBHP) and hydrogenated vegetable oil (HVO), commercially called Dalda, was bought from a local store.

Methods

Catalyst Preparation

To prepare geopolymer, AA and GP were thoroughly mixed for 10 min at two different proportions (w/w) of 1:2 and 1:3, and the final samples obtained were named G1 and G2, respectively. The viscous pastes were set for geopolymerization in covered plastic molds for 6 h to prevent evaporation followed by uncovered drying at 40°C overnight.

The geopolymer was modified with reactive emulsion templating, ion-exchange, and sulfation treatment. For reactive emulsion templating, HVO was chosen as a fatty acid source and was employed to assimilate additional porosity in geopolymer. In this treatment, HVO was added to the mixture of AA and GP at a 1:1 weight ratio of alkali activator to HVO. The addition of HVO to alkaline AA initiated an in-situ saponification reaction between fatty acids of vegetable oil and the alkali activator. It produced fatty acid salts or soap in the geopolymer, which occupied the structure (pores) of the geopolymer (Medpelli, Seo, and Seo 2014) (Butler, Hopkinson, and Cooper 2003). The mixture was stirred for 15 minutes and then cured for 24 h at 60°C. After curing, the solid mass was crushed to powder form and repeatedly washed with hot distilled water to remove the water-soluble soaps formed during saponification. The porous material thus produced was named as G1_OIL.

To beef up the surface properties of the G1_OIL sample further, the sample was ion-exchanged with NH_4Cl as per previous reports (O'Connor et al. 2010). In this process, 1 g of sample was vigorously stirred in 100 mL of 0.1 M NH_4Cl for 12 hours to facilitate the exchange of charge balancing Na^+ ions with NH_4^+. The powder was then removed from the solution via vaccum filtration and cleaned to clear out the remaining alkali and NH_4^+ ions. The sample was then dried overnight at 40°C, and calcined at 550°C for 15 min for the decomposition of NH_4^+ to H^+ (Alzeer and MacKenzie 2018). The resultant powdered sample was named G1_OIL_NH3.

To facilitate better interaction between basic DBT and the catalyst, the G1_OIL_NH3 sample was modified via treatment with sulfuric acid. For this, a weighed amount of the geopolymer was dipped in 30 mL of 1 M H_2SO_4 solution to impregnate sulfate ions in the geopolymer which enhanced its acidic properties by imposing super-acidity on the ion-exchanged geopolymer. The Hammett acidity function (H_o) of sulfuric acid (−12) indicated that it is a superacid and can intensify the acidic properties of ion-exchanged geopolymer. After 30 minutes, the remaining water was vaporised and the resultant residue was dried for 12 h. After complete drying of the residue, the sample was calcined for 4 h at 650°C. This sulfated form of geopolymer was named G1_OIL_NH3_S. During the water removal and drying step, impregnation of sulfate into the geopolymer surface probably followed the reaction stated below (Kumari et al. 2021),

$$\text{-(HO-Si-O-Al-OH)-}_n + xH_2SO_4 \longrightarrow \text{-(OH-Si-O-Al-O)}_{n-2x}(SO_4)_2 + 2xH_2O$$

A plausible structure of sulfated geopolymer is given in our previous paper (Kumari et al. 2021).

Experimental Procedure

The oxidation reaction was carried out in a three-necked glass batch reactor equipped with a condenser and placed in a water bath maintained at 50^0C whose schematic diagram can be found in the literature (Maity et al. 2012). The model fuel containing 500 ppm of S was prepared by dissolving appropriate amount of DBT in isooctane. 0.5 g of catalyst, 40 mL of model fuel and an appropriate amount of TBHP corresponding to the oxidant to sulfur ratio of 10:1 were fed to the reactor and the reaction was continued for 90 minutes with a stirring speed of 1000 rpm. Reaction samples (40 μL) were collected at 10 min intervals, filtered to remove catalyst particles and analyzed in HPLC [Perkin Elmer, Series 200] equipped with a reversed-phase Agilent SB C-18 column and a Perkin Elmer Series 200UV/VIS detector set at 254 nm with 90% methanol in water as the mobile phase.

Catalyst Characterization

Catalysts were characterized by X-ray Fluorescence Spectroscopy (XRF), (Panalytical, Axios, Netherland), BET surface area analysis (AS1 MP/Chemi-LP, USA), Scanning Electron Micrography (SEM), (Zeol and Zeiss with Oxford EDS Detector), X-ray Diffraction (XRD), (Panalytical 3050/60), Energy-Dispersive X-ray Spectroscopy (EDS) (Oxford EDS Detector), Analytical Transmission Electron Microscopy (TEM), (Tecnai G2), NH_3-Temperature Programmed Desorption (NH_3-TPD) (Quantachrome ChemBET 3000) analysis, and Fourier Transform Infrared Spectroscopy (FTIR) (Perkin Elmer Spectrum-100 FTIR) analysis.

Results and Discussion

Compositional Analysis

The compositional analysis of GP and AA was done by XRF and reported in our previous paper (Kumari et al. 2021). The composition of the GP resembled that of a fly ash material, indicating that the precursor can possibly be of fly ash origin. Kiran Global Green Cement reported that the AA was prepared from quartz and soda ash in the ratio of 2.0 ± 0.1 using the furnace route. The analysis also revealed that the AA contained sodium, indicating that the prepared geopolymer will be a sodium one.

SEM Analysis

The SE micrographs of GP, synthesized geopolymers (G1 and G2), and the modified geopolymers (G1_OIL, G1_OIL_NH3, G1_OIL_NH3_S) is shown in Figure 1. The SE micrograph of GP shows spherical-shaped particles, which are similar to the morphology of geopolymers prepared from fly ash sources, and was previously reported (Alzeer and MacKenzie 2018). Hence, it may be assumed that the present commercial GP may have fly ash origin. The SEM images of G1 and G2 show the presence of mesopores in the range of 10 to 40 nm, and this observation is similar to the geopolymer obtained by Papa et al. (Papa et al. 2018). SEM image of G1_OIL sample shows uniformly distributed mesopores and some macropores of the size range of 56 to 76 nm. The formation of macropores in the sample was possibly due to the entrapment of bubbles in the geopolymer matrix formed by the saponification reaction between HVO and AA. The macropores can be expected to reduce the restrictions on the diffusion of bulky DBT from porous channels to the acidic active centers present inside the catalyst. The presence of well-dispersed nanoparticles in the size range of 20 to 70 nm in G1_OIL_NH3 is also confirmed by their SEM images (Figure 1). This sample also contains mesopores ranging between 17 to 26 nm. After sulfation treatment, macropores in the size ranges of 70 to 128 nm appeared in the G1_OIL_NH3_S sample (Figure 1) which proves that macropores weren't choked by the bulky sulfate ions after sulfation treatment, thereby keeping the inner active centers accessible to the reactants.

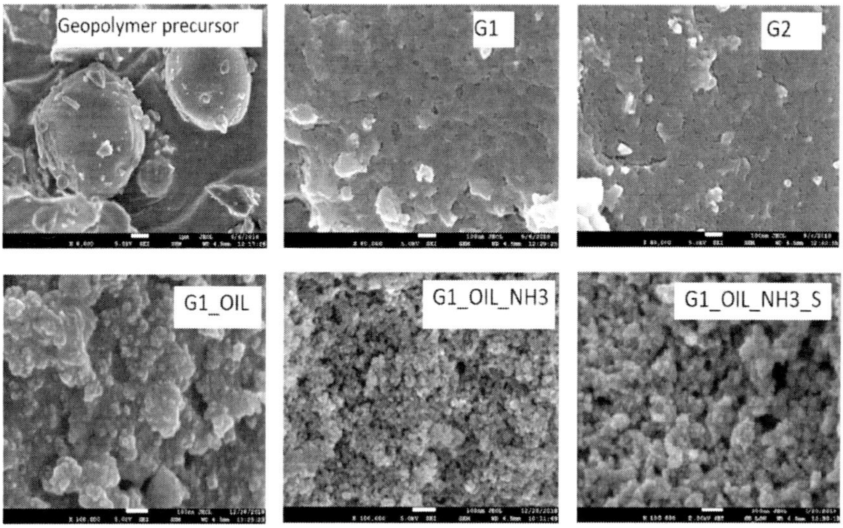

Figure 1. SEM images of GP, G1, G2 and the modified geopolymers.

TEM Analysis

SEM analysis revealed the formation of nanoparticles into the ion-exchanged sample and the same was confirmed by the TEM image given in Figure 2. The

TEM image shows cylindrical and spherical-shaped nanoparticles in a size range of 10-20 nm on the surface. These nanoparticles are expected to have helped in increasing the surface area of the catalyst.

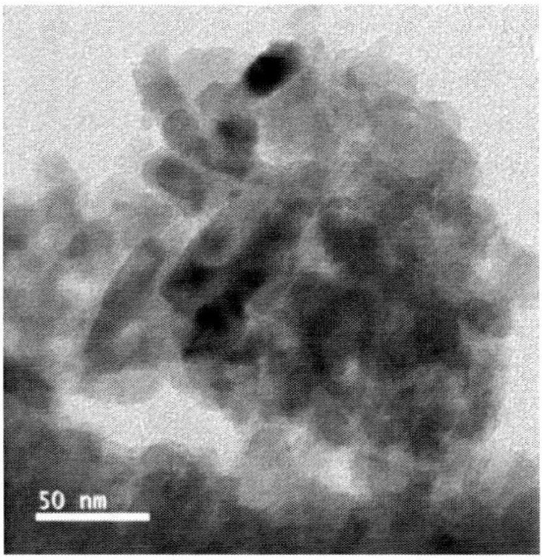

Figure 2. TEM image of G1_OIL_NH3 sample.

BET Analysis

The surface area, average pore diameter and pore volume of the GP, geopolymer G1, G2 and modifed geopolymers is reported in Table 1. It shows that after alkali treatment, the average pore size of the prepared geopolymer increased significantly from 7.2 (GP) to 15.8 nm (G1). This increase could be attributed to the dissolution of pore walls into alkali solution, as a result of which the surface area also decreased from 8.55 (GP) to 4.5 m^2/g (G1) (Choi et al. 2006) (Alzeer, MacKenzie, and Keyzers 2017). The dissolution of pore walls also resulted in an increase in the pore volume, which was higher in case of G1 geopolymer. Thus, the ratio of GP and AA used for the prepration of G1 geopolymer was chosen for emulsion templating. The geopolymer G1 was modified via emulsion templating, ion-exchange and sulfation treatments. After reactive emulsion treatment, the surface area of the G1_OIL sample significantly reduced to 1.34 m^2/g, but the pore size increased to 18 nm. The decrement in the surface area was expected to be the consequence of larger pores formed due to the soap bubbles generated from soap formed by saponification reaction of vegetable oil and water, and subsequent removal of entrapped emulsion resulted in porous structure (Maity et al. 2012). After this step, a decrease in the pore volume was noticed, which was possibly due to entrapment of some excess oil and soap products into its pores. In the next step, the G1_OIL sample was ion-exchanged with NH_4Cl to produce G1_OIL_NH3. The surface area increased significantly from 1.34 to 71.54 m^2/g due to the presence of nanoparticles uniformly distributed on its surface. The same was confirmed by theSEM images shown in Figure 1 too. The nanoparticles thus formed were found to clog the pores of the G1_OIL_NH3 sample, possiblydue to which a decrease in the average pore radius of G1_OIL_NH3 from 18 to 17.2 nm was observed which is shown in Table 1. Afterwards, when the G1_OIL_NH3 sample was undergone through sulfation process to produce G1_OIL_NH3_S, the surface area was reduced from 71.54 to 20.5 m^2/g, due to the addition of sulfate ions in the bulk phase of geopolymer (Choi et al. 2006). This decrement in the surface area could be due to choking of selected mesopores during the sulfation step. On the contrary, an increase in the average pore diameter from 17.2 to 18.7 nm was noticed (Table 1). This increase was possibly due to washing off of some geopolymer mass leading to the disintegration of pore walls at some places. This increase was also noticeable from the increase in the pore volume of the final geopolymer in comparison to the oil treated geopolymer G1_OIL.

Table 1. Surface area, average pore diameter and pore volume of GP, G1, G2 and the modified geopolymers

Sample	Surface area (m^2/g)	Average pore diameter (nm)	Pore volume (cc/g)
GP	8.55	7.2	0.014
G1	4.5	15.8	0.025
G2	6.47	10.0	0.019
G1_OIL	1.34	18.0	0.006
G1_OIL_NH3	71.54	17.2	0.310
G1_OIL_NH3_S	20.5	18.7	0.096

Ammonia Temperature Programmed Desorption (NH_3-TPD) Analysis

Table 2 shows the concentration of acidic active sites of GP, G1, G2, and the modified geopolymer catalysts (G1_OIL, G1_OIL_NH3, G1_OIL_NH3_S) analyzed in NH_3-TPD. It is observed that the GP sample possessed an insignificant concentration of acidic sites, while it was comparatively higher for geopolymers G1 and G2. Compared to the geopolymers prepared by

Alzeer et al. (Alzeer, MacKenzie, and Keyzers 2016), our geopolymer contained a greater number of acidic sites. During emulsion templating and ion exchange treatment, the concentration increased steadily to 9.83 mmol/g, which can be expected to have occurred due to partial dealumination during the ion-exchange treatment step, which gave rise to extra framework aluminum, a source of acidic sites in crystalline aluminosilicates (zeolites) (Alzeer and MacKenzie 2018). After treatment with sulfuric acid, the concentration of active sites jumped to 46.99 mmol/g, indicating that the superacid had a notable influence on the properties of the geopolymer and could successfully impart super acidic properties onto the parent geopolymer. This gradual and steep increase in the acidic properties of the geopolymer was expected to aid DBT conversion significantly.

Table 2. Concentration of acidic sites in GP, G1, G2 and the modified geopolymers

Sample	Acidic site concentration (mmol/g)
GP	0.16
G1	1.71
G2	1.55
G1_OIL	4.01
G1_OIL_NH3	9.83
G1_OIL_NH3_S	46.99

EDS Analysis

The elemental analysis of the geopolymer samples is tabulated in Table 3. The EDS findings of GP were coherent with the XRF results given in our previous paper (Kumari et al. 2021). It is noticed that the major elements present in GP were aluminum, silicon and calcium.. After alkali activation, the presence of sodium originating from the alkali activator in all the samples was confirmed. Post emulsion templating treatment, the percentage of sodium in the sample was reduced, indicating that some of the sodium was consumed in the saponification reaction to produce soap and glycerol. These soap molecules were entrapped in the geopolymer matrix and led to a macroporous geopolymer structure after removal. After ion-exchange treatment, the sodium content was further reduced, indicating a successful ion exchange with ammonium ions. Finally, sulfur could be seen in high concentrations after sulfation treatment, indicating a successful impregnation of sulfate ions onto the geopolymer.

Table 3. EDS analysis of geopolymer and the modified geopolymers

Sample	Carbon	Oxygen	Magnesium	Aluminum	Silicon	Potassium	Calcium	Sodium	Sulfur
GP	8.06	42.00	3.35	9.23	15.02	0.27	22.07	-	-
G1	10.09	24.00	2.44	7.10	18.23	0.28	22.36	15.5	-
G2	7.09	27.43	3.49	7.35	16.54	0.30	20.80	17.00	-
G1_OIL	52.03	27.78	0.41	1.43	11.22	-	2.22	4.91	-
G1_OIL_NH3	26.32	50.92	1.06	3.48	10.87	-	5.30	2.05	-
G1_OIL_NH3_S	2.99	57.04	1.67	4.90	7.59	-	5.55	1.17	19.09

XRD Analysis

XRD of GP, G1, G2, and the modified geopolymers is shown in Figures 3(a) and 3(b). The XRD of the GP sample showed the reflection of kyanite, quartz and aluminum silicate at 2θ values of 20.77, 38.22, and 44.77° respectively. Two reflections of mullite were also visible at 2θ of 26.57 and 50.17°. These observations confirmed that the GP sample was rich in silicon and aluminum and could act as a raw material for the preparation of geopolymer. The diffractogram of both the geopolymers G1 and G2 were similar. They possessed a characteristic broad hump between 20-40° 2θ which is typically present in a well-formed geopolymer (Alzeer and MacKenzie 2018). Along with the broad hump, a sodium aluminosilicate peak was also seen at 30.17° 2θ. The kyanite, mullite and quartz peaks in the GP sample could also be seen in both the geopolymer samples G1 and G2.

The three-staged treatment brought significant crystallographic changes in the geopolymer, which their corresponding diffractograms could apprehend. Reactive emulsion templating was done in-situ with the geopolymer formation. The emulsion templated geopolymer sample possessed the characteristic broad hump between 20-40° 2θ, a sharp sodium aluminosilicate peak at 26.87° 2θ, and several small sodium aluminosilicate peaks at 21.12, 31.07, 35.27, 39.67, 41.12 and 50.37° 2θ. The quartz, kyanite, and mullite peaks in the GP were absent in the emulsion templated geopolymer, indicating that most of the aluminum and silicon present in the precursor had reacted with the AA and converted into geopolymer. After ion-exchange treatment, the broad hump remained intact with the sodium aluminosilicate peaks at 26.87, 31.07, 50.37 and 60.27° 2θ. Apart from this, no new crystal phases appeared in the ion-exchanged geopolymer. Sulfation treatment significantly affected the crystallographic properties of the geopolymer, and a number of sodium aluminosilicate peaks appeared at 15.12, 21.12, 25.22, 26.87, 30.17, and 32.93° 2θ.

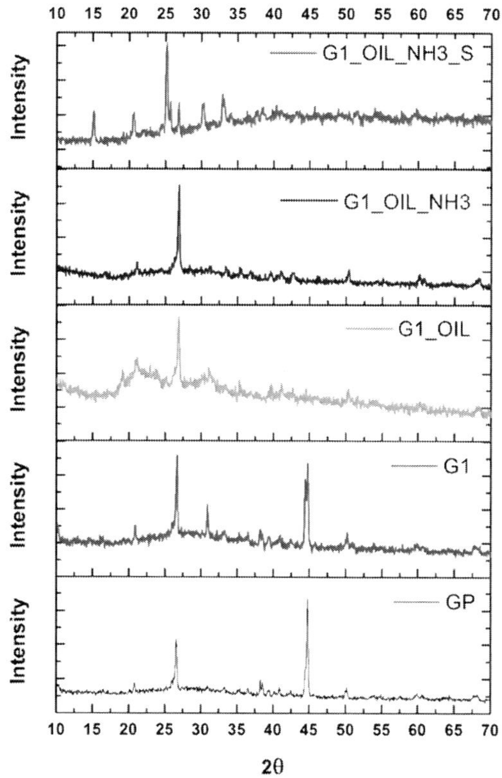

Figure 3. XRD peak analysis of GP, G1 and modified geopolymers.

The XRD analysis results were used to determine the crystal size and degree of crystallinity. They were calculated based on the highest XRD diffraction peak using the following equations:

Crystallite size

$$d = \frac{K\lambda}{\beta cos\theta} \quad (1)$$

where;

d: Crystallite size in nm

K: Scherrer Constant = 0.89

λ: X-ray wavelength = 1.5417Å

β: FWHM (Full width at half maximum)

θ: XRD peak position (one half of 2θ)

$$\text{Degree of crystallinity} = \frac{Area\ of\ total\ crystalline\ peaks}{Area\ of all\ peaks\ (amorphous\ \&\ crystalline)} \tag{2}$$

It is noticed that the crystal size of GP was 37 nm, which gradually decreased to 17.45 nm (G1_OIL_NH3_S) after treatment. Table 4 shows that GP possesses a low crystalline phase (30.83%) which increased appreciably to 76.68% for the sulfated geopolymer (G1_OIL_NH3_S) sample. The modified geopolymer had a greater crystallinity which helped it to acquire a huge number of active sites for enhancing the reaction rate (Mohammadparast, Halladj, and Askari 2015) (Pirutko et al. 1996) (Othman et al. 2010). It was found that, with increasing degree of crystallinity, several specified crystal phases were promoted to participate in the reaction, which ultimately enhanced the DBT conversion (Oliver et al. 2004) (Nicolaides, Sincadu, and Scurrell 2002).

Table 4. Crystal size and degree of crystallinity of GP, G1 and the modified geopolymers

Sample	Crystal size (nm)	Degree of crystallinity (%)
GP	37	30.83
G1	35.27	37.69
G1_OIL	29.92	54.72
G1_OIL_NH3	25.16	65.66
G1_OIL_NH3_S	17.45	76.68

FTIR Analysis

Figure 4 shows the FTIR spectra of the modified geopolymers G1_OIL, G1_OIL_NH3 and G1_OIL_NH3_S prepared after each stage of the employed treatment. The oil templated sample (G1_OIL) shows a sharp band at around 1045 cm^{-1} attributed to the stretching vibration of the Si-O-Al group, a typical characteristic of a geopolymer sample (Alzeer and MacKenzie 2018).

It also contains small shoulders at around 1167, 924,750, 693, and 550 cm^{-1}, typically ascribed to Si-O-Si stretch, Si-OH bend, Al-O stretch, bending vibration of Si-O tetrahedra, and Si-O-Al symmetric stretch, respectively (Alzeer and MacKenzie 2018) (Z. Wang and Lu 2020). All of these bands are typically seen in geopolymer samples. The sample also displayed bands in the region of 1400 cm^{-1}, appearing due to the presence of carbonate species formed due to atmospheric carbonation of geopolymer (Alzeer and MacKenzie 2018). The atmospheric carbonation also resulted in a small peak at 2340 cm^{-1}, attributed to CO_2 asymmetric stretching vibration (Alzeer and MacKenzie 2018). Two small but sharp bands appearing at 2843 and 2916 cm^{-1} are ascribed to C–H symmetric and asymmetric stretching vibrations, and the one at around 1550 cm^{-1} is attributed to the anti-symmetric stretching vibration of the carboxyl group ($-COO^-$) appearing due to the presence of some amount of organic phase in the sample (Z. Wang and Lu 2020). A broad band at about 3400 cm^{-1} is ascribed to H-bonded silanol nests (Alzeer and MacKenzie 2018).

The ion-exchanged sample (G1_OIL_NH3) displayed a sharp peak at around 1060 cm^{-1}, typically ascribed to the Si-O-Al stretch of geopolymer samples (Alzeer and MacKenzie 2018). The small shift in the band compared to the oil treated sample indicated that the treatment had a strong effect on the properties of the geopolymer. The sample also displayed a band at around 694 cm^{-1} that can be ascribed to bending vibrations of Si-O tetrahedral (Z. Wang and Lu 2020). The small shoulder at around 800 cm^{-1} can be ascribed to the presence of octahedral aluminum species (EFAl entities), which act as Lewis acid sites (Alzeer and MacKenzie 2018). Unlike the G1_OIL sample, the G1_OIL_NH3 sample showed a small shoulder at 1640 cm^{-1}, typically associated with H-OH stretching mode of adsorbed water (Alzeer and MacKenzie 2018). Similar to the G1_OIL sample, the G1_OIL_NH3 sample also displayed bands in the region of 2800-2900 cm^{-1} attributed to C–H symmetric and asymmetric stretching vibrations (Z. Wang and Lu 2020). However, the reduced intensity of these bands indicates the successful removal of the template. A band at around 1561 $cm-^{1}$ is generally attributed to the anti-symmetric stretching vibration of the carboxyl group ($-COO^-$). It can also be associated with the presence of Bronsted acid sites (Z. Wang and Lu 2020). Presence of the band confirms the presence of H-bonded silanol nests in the G1_OIL_NH3 sample at 3440 cm^{-1}. However, the shrinkage of the band indicates partial destruction of the silanol groups due to thermal treatment.

The sulfated sample (G1_OIL_NH3_S) displayed a sharp peak and two shoulders in the range of 1000-1200 cm^{-1}, which can be ascribed to Si-O-Al stretch and the association of geopolymer network with bidentate sulfate ions and the bands in the range of 600 cm^{-1} can be attributed to the bending vibrations of Si-O tetrahedra (J. R. Sohn and Kim 1989) (Mishra, Tyagi, and Jasra 2003) (Yamaguchi 1990) (Z. Wang and Lu 2020). The sample also displayed a sharp band at around 1640 cm^{-1}, typically ascribed to the bending vibrations of chemisorbed water molecules associated with the sulfate groups (B. Wang, Zhu, and Ma 2009). A small band appearing at around 2500 cm^{-1} can be ascribed to the asymmetric stretching mode of CO_2 resulting due to atmospheric carbonation of geopolymer (Alzeer, MacKenzie, and Keyzers 2017). Unlike G1_OIL and G1_OIL_NH3 samples, the sharp band corresponding to C–H symmetric and asymmetric stretching vibrations was absent in the sulfated geopolymer sample. The broad peak at around 3400 cm^{-1} confirmed the presence of H-bonded silanol nests in the G1_OIL_NH3_S sample.

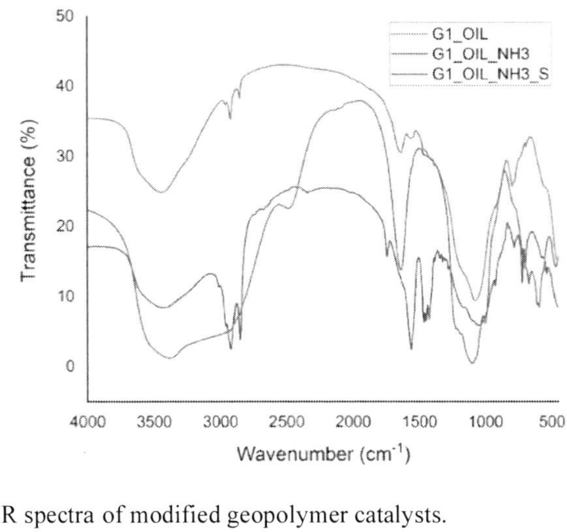

Figure 4. FTIR spectra of modified geopolymer catalysts.

Catalytic Activity Test of Geopolymers in DBT Oxidation Reaction

The catalytic activity of GP, G1, G2, and the modified geopolymers was studied by subjecting the catalyst to a DBT oxidation reaction. The reaction conditions are specified in the experimental section. Figures 5(a) and 5(b) reveal a gradual increase in the conversion of DBT from 39 to 49%, then from 49 to 54%, and finally from 54 to 74% when G1, G1_OIL, G1_OIL_NH3, and G1_OIL_NH3_S catalysts were used respectively. As mentioned earlier, the

G1_OIL catalyst possesses a much bigger pore size compared to G1, which causes the DBT molecules to gain better access to the active sites, in turn, to face lesser diffusion resistance, and this ultimately enhances the conversion of DBT (Chang and Crynes 1986) (Li et al. 2017). The conversion was observed to increase further from 49 to 54%, and the increased acidity of G1_OIL_NH3 can well explain this increase. The basic characteristic of DBT has another major role in improving conversion (Yan et al. 2013).

The DBT conversion was further enhanced with G1_OIL_NH3 catalyst due to the presence of a higher concentration of active sites on its surface. The catalyst G1_OIL_NH3_S in the reaction results in a remarkable increase in the conversion from 54 to 74% due to the stronger intensity of acidic active sites in the G1_OIL_NH3_S catalyst. The strong acidity of the catalyst aided in better acid-base interaction between the acidic active sites of the catalyst and basic DBT, and thus despite having a smaller surface area compared to G1_OIL_NH3, the G1_OIL_NH3_S catalyst gave better DBT conversion.

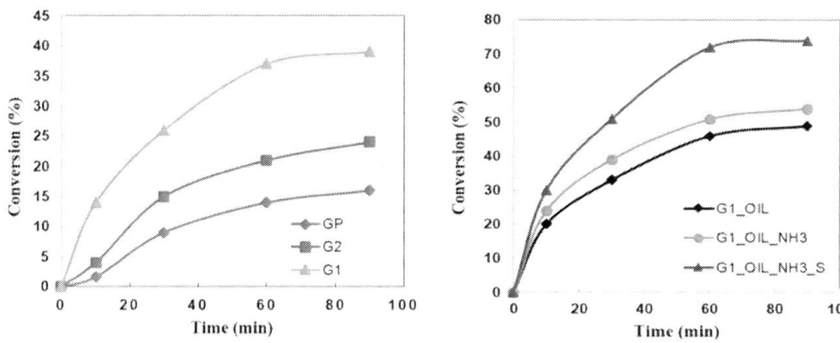

(a) (b)

Figure 5. Catalytic activity test of (a) GP, G1 and G2, (b) modified geopolymers in DBT oxidation.

Conclusion

A geopolymer prepared with fly ash-based precursor and sodium-containing alkali activator was modified via three-staged treatment, including reactive emulsion templating, ion-exchange, and sulfation steps. Emulsion templating treatment aided in introducing hierarchical porosity into the geopolymer and the generation of porosity was confirmed by the SE micrographs. Along with an increase in the surface area of the catalyst, the ion-exchange treatment helped the conversion of Na-form of geopolymer to its acidic H-form. The acidity of the geopolymer was further enhanced by subjecting the geopolymer to sulfation treatment. After the formulated three-staged treatment, a drastic improvement in the concentration of acidic active sites from 1.71 to 46.99 mmol/g was observed. As is evident from the XRD analysis, the treatment also had a significant effect on the degree of crystallinity of the geopolymer. The prepared geopolymer catalyst yielded 74% conversion of DBT and can thus be regarded as an efficient catalyst for pre-treatment of high sulfur fuel oils.

References

Akbari, Hamid, Robert Mensah-biney, and Jonathan Simms. (2015). Production of Geopolymer Binder from Coal Fly Ash to Make Cement-Less Concrete, 1–8. *Paper presented at the 2015 World of Coal Ash (WOCA) Conference*, Nasvhille, TN-May 5-7, 2015.

Alzeer, Mohammad I. M., and Kenneth J. D. MacKenzie. (2018). Synthesis and Catalytic Properties of New Sustainable Aluminosilicate Heterogeneous Catalysts Derived from Fly Ash. *ACS Sustainable Chemistry and Engineering* 6 (4): 5273–82. https://doi.org/10.1021/acssuschemeng.7b04923.

Alzeer, Mohammad I. M., Kenneth J. D. MacKenzie, and Robert A. Keyzers. (2016). Porous Aluminosilicate Inorganic Polymers (Geopolymers): A New Class of Environmentally Benign Heterogeneous Solid Acid Catalysts. *Applied Catalysis A: General* 524: 173–81. https://doi.org/10.1016/j.apcata.2016.06.024.

Alzeer, Mohammad I M, Kenneth J D MacKenzie, and Robert A Keyzers. (2017). Facile Synthesis of New Hierarchical Aluminosilicate Inorganic Polymer Solid Acids and Their Catalytic Performance in Alkylation Reactions. *Microporous and Mesoporous Materials* 241: 316–25. https://doi.org/10.1016/j.micromeso.2016.12.018.

Arata, Kazushi. (1990). Solid Superacids. In, edited by D D Eley, Herman Pines, and Paul B B T - *Advances in Catalysis Weisz*, 37:165–211. Academic Press. https://doi.org/10.1016/S0360-0564(08)60365-X.

Asim, Nilofar, Mohammad Alghoul, Masita Mohammad, Mohamad Hassan Amin, M Akhtaruzzaman, Nowshad Amin, and Kamaruzzaman Sopian. (2019). Emerging Sustainable Solutions for Depollution: Geopolymers. *Construction and Building Materials* 199: 540–48. https://doi.org/10.1016/j.conbuildmat.2018.12.043.

Bai, Chengying, and Paolo Colombo. (2018). Processing, Properties and Applications of Highly Porous Geopolymers: A Review. *Ceramics International* 44 (14): 16103–18. https://doi.org/10.1016/j.ceramint.2018.05.219.

Bazyari, Amin, Abbas A Khodadadi, Alireza Haghighat Mamaghani, Javad Beheshtian, Levi T Thompson, and Yadollah Mortazavi. (2016). Microporous Titania–Silica Nanocomposite Catalyst-Adsorbent for Ultra-Deep Oxidative Desulfurization. *Applied Catalysis B: Environmental* 180: 65–77. https://doi.org/10.1016/j.apcatb.2015.06.011.

Brus, Jiri, Sabina Abbrent, Libor Kobera, Martina Urbanova, and Pavel Cuba. (2016). Chapter Two - Advances in 27Al MAS NMR Studies of Geopolymers. In, edited by Graham A B T - *Annual Reports on N M R Spectroscopy* Webb, 88:79–147. Academic Press. https://doi.org/10.1016/bs.arnmr.2015.11.001.

Butler, R, I Hopkinson, and A I Cooper. (2003). Synthesis of Porous Emulsion-Templated Polymers Using High Internal Phase CO_2-in-Water Emulsions. *Journal of the American Chemical Society* 125 (47): 14473–81. https://doi.org/10.1021/ja037570u.

Chang, Hong Ju, and B. L. Crynes. (1986). Effect of Catalyst Pore and Pellet Sizes on Deactivation in SRC Oil Hydrotreatment. *AICHE Journal* 32 (2): 224–32.

Chen, Kun, Xiao-Min Zhang, Xian-Feng Yang, Meng-Gai Jiao, Zhen Zhou, Ming-Hui Zhang, Dan-Hong Wang, and Xian-He Bu. (2018). Electronic Structure of Heterojunction MoO_2/g-C_3N_4 Catalyst for Oxidative Desulfurization. *Applied Catalysis B: Environmental* 238: 263–73. https://doi.org/10.1016/j.apcatb.2018.07.037.

Chen, Yuchi, Xian Zhou, Sha Wan, Rui Zheng, Jun Tong, Haobo Hou, and Teng Wang. (2019). Synthesis and Characterization of Geopolymer Composites Based on Gasification Coal Fly Ash and Steel Slag. *Construction and Building Materials* 211: 646–58. https://doi.org/10.1016/j.conbuildmat.2019.03.292.

Choi, Dong Ho, Ji Won Park, Jong-Ho Kim, Yoshihiro Sugi, and Gon Seo. (2006). Liquid-Phase Degradation of HDPE over Alkali-Treated MFI Zeolites with Mesopores. *Polymer Degradation and Stability* 91 (12): 2860–66. https://doi.org/10.1016/j.polymdegradstab.2006.09.002.

Davidovits, J. (1991). Geopolymers. *Journal of Thermal Analysis* 37 (8): 1633–56. https://doi.org/10.1007/BF01912193.

Dezfouli H R, Rangaraju P R, Kothala V S K. (2018). Influence of Selected Parameters on Compressive Strength of Geopolymer Produced from Ground Glass Fiber. *Construction and Building Materials* 162: 393–405.

Keppert, Martin, Eva Vejmelková, Petr Bezdi, Magdaléna Dole, Č Monika, Lenka Scheinherrová, Jaroslav Pokorný, Martin Vy, Pavla Rovnaníková, and Č Robert. (2018). Red-Clay Ceramic Powders as Geopolymer Precursors: Consideration of Amorphous Portion and CaO Content. *Applied Clay Science* 161: 82–89.

Kovářík, Tomáš, Tomas Krenek, M Pola, David Rieger, Jaromir Kadlec, and P Franče. (2017). Ceramic-like Open-Celled Geopolymer Foam as a Porous Substrate for Water Treatment Catalyst. *IOP Conference Series: Materials Science and Engineering* 175 (February): 12044. https://doi.org/10.1088/1757-899X/175/1/01 2044.

Kumari, Snehlata, Biswajit Saha, Debdatta Guha, and Sonali Sengupta. (2021). Sequential Post-Synthetic Modification and Evaluation of Catalytic Activity of Hierarchically Porous Sulfated Geopolymer in the Oxidative Desulfurization of Dibenzothiophene. *Catalysis Surveys from Asia*. https://doi.org/10.1007/s10563-021-09328-6.

Li, Hansheng, Jungang Wang, Congbiao Chen, Litao Jia, Bo Hou, and Debao Li. (2017). Effects of Macropores on Reducing Internal Diffusion Limitations in Fischer–Tropsch Synthesis Using a Hierarchical Cobalt Catalyst. *RSC Adv.* 7 (16): 9436–45. https://doi.org/10.1039/C6RA27166A.

Ma, Wenwen, Yan Xu, Kaiwen Ma, and Hong Zhang. (2016). Electrospinning Synthesis of $H_3PW_{12}O_{40}/TiO_2$ Nanofiber Catalytic Materials and Their Application in Ultra-

Deep Desulfurization. *Applied Catalysis A: General* 526: 147–54. https://doi.org/10.1016/j.apcata.2016.08.021.

Maity, Uttam, Rosilda Selvin, Jayanta Basu, and Sonali Sengupta. (2012). Application of Synthesized Nano-Crystalline Titanium Silicate-1 in the Oxidation of Thiophene in a Kinetic Approach. *Journal of Nanoengineering and Nanomanufacturing* 2 (September): 241–47. https://doi.org/10.1166/jnan.2012.1079.

Medpelli, Dinesh, Jung-min Seo, and Dong-Kyun Seo. (2014). Geopolymer with Hierarchically Meso-/Macroporous Structures from Reactive Emulsion Templating. *Journal of the American Ceramic Society* 97 (January). https://doi.org/10.1111/jace.12724.

Metz, Bert, Ogunlade Davidson, Heleen de Coninck, Manuela Loos, and Leo Meyer. (2005). *IPCC*, 2005. Cambridge CB2 2RU ENGLAND. https://www.ipcc.ch/report/carbon-dioxide-capture-and-storage/.

Mishra, Manish. K, Beena Tyagi, and Raksh. V Jasra. (2003). Effect of Synthetic Parameters on Structural, Textural, and Catalytic Properties of Nanocrystalline Sulfated Zirconia Prepared by Sol−Gel Technique. *Industrial & Engineering Chemistry Research* 42 (23): 5727–36. https://doi.org/10.1021/ie030099t.

Mohammadparast, Farshid, Rouein Halladj, and Sima Askari. (2015). The Crystal Size Effect of Nano-Sized ZSM-5 in the Catalytic Performance of Petrochemical Processes: A Review. *Chemical Engineering Communications* 202 (4): 542–56. https://doi.org/10.1080/00986445.2014.952815.

Nicolaides, C P, N P Sincadu, and M S Scurrell. (2002). NAS (Novel Aluminosilicates) as Catalysts for the Aromatisation of Propane: Studies of Zinc and Gallium Modified Zeolite-Based Systems Having Various Extents of XRD Crystallinity. *Catalysis Today* 71 (3): 429–35. https://doi.org/10.1016/S0920-5861(01)00470-9.

O'Connor, Sean J, Kenneth J D MacKenzie, Mark E Smith, and John V Hanna. (2010). Ion Exchange in the Charge-Balancing Sites of Aluminosilicate Inorganic Polymers. *J. Mater. Chem.* 20 (45): 10234–40. https://doi.org/10.1039/C0JM01254H.

Oliver, J M, J M López Nieto, P Botella, and A Mifsud. (2004). The Effect of PH on Structural and Catalytic Properties of MoVTeNbO Catalysts. *Applied Catalysis A: General* 257 (1): 67–76. https://doi.org/10.1016/S0926-860X(03)00632-X.

Othman, Siti, Suraya Abdul Rashid, Tinia Mohd Ghazi, and Norhafizah Abdullah. (2010). Effect of Postdeposition Heat Treatment on the Crystallinity, Size, and Photocatalytic Activity of TiO_2 Nanoparticles Produced via Chemical Vapour Deposition. *Journal of Nanomaterials* 2010 (June). https://doi.org/10.1155/2010/ 512785.

Papa, Elettra, Valentina Medri, Smail Amari, Jérémie Manaud, Patricia Benito, Angelo Vaccari, and Elena Landi. (2018). Zeolite-Geopolymer Composite Materials: Production and Characterization. *Journal of Cleaner Production* 171: 76–84. https://doi.org/10.1016/j.jclepro.2017.09.270.

Pirutko, L V, K A Dubkov, L P Solovyeva, and G I Panov. (1996). Effect of ZSM-11 Crystallinity on Its Catalytic Performance in Benzene to Phenol Oxidation with Nitrous Oxide. *Reaction Kinetics and Catalysis Letters* 58 (1): 105–10. https://doi.org/10.1007/BF02071112.

Saha, Biswajit, Sandeep Kumar Yadav, and Sonali Sengupta. (2018). Synthesis of Nano-Hap Prepared through Green Route and Its Application in Oxidative Desulfurisation. *Fuel* 222: 743–52. https://doi.org/10.1016/j.fuel.2018.02.139.

Sazama, Petr, Oleg Bortnovsky, Jiří Dědeček, Zdenka Tvarůžková, and Zdeněk Sobalík. (2011). Geopolymer Based Catalysts—New Group of Catalytic Materials. *Catalysis Today* 164 (1): 92–99. https://doi.org/10.1016/j.cattod.2010.09.008.

Sharma, Sudhanshu, Dinesh Medpelli, Shaojiang Chen, and Dong-Kyun Seo. (2015). Calcium-Modified Hierarchically Porous Aluminosilicate Geopolymer as a Highly Efficient Regenerable Catalyst for Biodiesel Production. *RSC Adv.* 5 (July). https://doi.org/10.1039/C5RA01823D.

Sohn, Jong, and Si Lee. (2004). Acidic Properties of Nickel Sulfate Supported on TiO_2–ZrO_2 and Catalytic Activity for Acid Catalysis. *Applied Catalysis A: General* 266 (July): 89–97. https://doi.org/10.1016/j.apcata.2004.01.034.

Sohn, Jong Rack, and Hae Won Kim. (1989). Catalytic and Surface Properties of ZrO_2 Modified with Sulfur Compounds. *Journal of Molecular Catalysis* 52 (3): 361–74. https://doi.org/10.1016/0304-5102(89)85045-X.

Wan, Qian, Feng Rao, and Shaoxian Song. (2017). Reexamining Calcination of Kaolinite for the Synthesis of Metakaolin Geopolymers - Roles of Dehydroxylation and Recrystallization. *Journal of Non-Crystalline Solids* 460: 74–80. https://doi.org/10.1016/j.jnoncrysol.2017.01.024.

Wang, Bo, Jianpeng Zhu, and Hongzhu Ma. (2009). Desulfurization from Thiophene by SO_4^{2-}/ZrO_2 Catalytic Oxidation at Room Temperature and Atmospheric Pressure. *Journal of Hazardous Materials* 164: 256–64. https://doi.org/10.1016/j.jhazmat.2008.08.003.

Wang, Zhuangzhuang, and Duyou Lu. (2020). Study on the Effect of Emulsifiers on the Pore Structures of Geopolymer Prepared by Emulsion Templating. *Materials Research Express* 7 (5): 55508. https://doi.org/10.1088/2053-1591/ab94f9.

Yamaguchi, Tsutomu. (1990). Recent Progress in Solid Superacid. *Applied Catalysis* 61 (1): 1–25. https://doi.org/10.1016/S0166-9834(00)82131-4.

Yan, Xue-Min, Ping Mei, Lin Xiong, Lin Gao, Qifeng Yang, and Lunjun Gong. (2013). Mesoporous Titania–Silica–Polyoxometalate Nanocomposite Materials for Catalytic Oxidation Desulfurization of Fuel Oil. *Catal. Sci. Technol.* 3 (8): 1985–92. https://doi.org/10.1039/C3CY20732C.

Zhang, Yao Jun, Ke Zhang, Le Kang, and Li Zhang. (2017). A New In_2O_3 and NiO Co-Loaded Fly Ash-Based Nanostructural Geopolymer for Photocatalytic H_2 Evolution. *Integrated Ferroelectrics* 182 (1): 1–9. https://doi.org/10.1080/10584587.2017.1352376.

Zhang, Yaojun, and Licai Liu. 2013. Fly Ash-Based Geopolymer as a Novel Photocatalyst for Degradation of Dye from Wastewater. *Particuology* 11 (3): 353–58. https://doi.org/10.1016/j.partic.2012.10.007.

Zhao, Rijie, Jianlong Wang, Dongdong Zhang, Yahui Sun, Baixin Han, Nan Tang, Na Wang, and Kaixi Li. (2017). Biomimetic Oxidative Desulfurization of Fuel Oil in Ionic Liquids Catalyzed by Fe (III) Porphyrins. *Applied Catalysis A: General* 532: 26–31. https://doi.org/10.1016/j.apcata.2016.12.008.

Bibliography

3D concrete printing technology: construction and building applications

LCCN	2018964172 s
Type of material	Book
Main title	3D concrete printing technology: construction and building applications / edited by Jay G Sanjayan, Centre for Sustainable Infrastructure, Faculty of Science, Engineering and Technology, Swinburne University of Technology, Hawthorn, VIC, Australia, Ali Nazari, Centre for Sustainable Infrastructure, Faculty of Science, Engineering and Technology, Swinburne University of Technology, Hawthorn, VIC, Australia, Behzad Nematollahi, Centre for Sustainable Infrastructure, Faculty of Science, Engineering and Technology, Swinburne University of Technology, Hawthorn, VIC, Australia.
Published/Produced	Oxford [England]; Cambridge, MA: Butterworth-Heinemann, [2019]
Description	xxiii, 425 pages: illustrations; 23 cm
ISBN	9780128154816
LC classification	TS171.95 .A143 2019
Related names	Sanjayan, Jay G., editor. Nazari, Ali (Materials scientist), editor. Nematollahi, Behzad, editor.
Contents	3D Concrete Printing for Construction Applications / Jay G. Sanjayan and Behzad Nematollahi - Performance-Based Testing of Portland Cement Concrete for Construction-Scale 3D Printing / Ali Kazemian, Xiao Yuan, Ryan Meier and Behrokh Khoshnevis - Building Applications Using Lost Formworks Obtained Through Large-Scale Additive Manufacturing of Ultra-High-Performance Concrete / N. Gaudillière, R. Duballet, C. Bouyssou, A. Mallet, Ph. Roux, M.

Zakeri and J. Dirrenberger - Fiber-Reinforced Cementitious Composite Design with Controlled Distribution and Orientation of Fibers Using Three-Dimensional Printing Technology / Young Jun Nam, Young Kwang Hwang, Ji Woon Park and Yun Mook Lim - Properties of 3D-Printed Fiber-Reinforced Portland Cement Paste / Manuel Hambach, Matthias Rutzen and Dirk Volkmer - 3D Concrete Printing: Machine Design, Mix Proportioning, and Mix Comparison Between Different Machine Setups / Zeina Malaeb, Fatima AlSakka and Farook Hamzeh - Investigation of Concrete Mixtures for Additive Construction / Todd S. Rushing, Peter B. Stynoski, Lynette A. Barna, Ghassan K. Al-Charr, Jedadiah F. Burroughs, Jameson D. Shannon, Megan A. Kreiger and Michael P. Case - Method for the Enhancement of Buildability and Bending Resistance of Three-Dimensional-Printable Tailing Mortar / Zhijian Li, Li Wang and Guowei Ma - Mechanical Properties of Structures 3D-Printed With Cementitious Powders / Peng Feng, Xinmiao Meng, Jian-Fei Chen and Lieping Ye - Factors Influencing the Mechanical Properties of Three-Dimensional Printed Products From Magnesium Potassium Phosphate Cement Material / Xiangpeng Cao and Zongjin Li - Development of Powder-Based 3D Concrete Printing Using Geopolymers / Ming Xia, Behzad Nematollahi and Jay G. Sanjayan - Interlayer Strength of 3D Printed Concrete: Influencing Factors and Method of Enhancing / Taylor Marchment, Jay G. Sanjayan, Behzad Nematollahi and Ming Xia - Properties of Powder-Based 3D Printed Geopolymers / Ming Xia, Behzad Nematollahi and Jay G. Sanjayan - Design 3D Printing Cementitious Materials Via Fuller Thompson Theory and Marson-Percy Model / Yiwei Weng, Mingyang Li, Ming Jen Tan snd Shunzhi Qian - Towards the Formulation of Robust

	and Sustainable Cementitious Binders for 3D Additive Construction by Extrusion / Dale P. Bentz, Scott Z. Jones, Isaiah R. Bentz and Max A. Peltz - Studying the Printability of Fresh Concrete for Formwork-Free Concrete Onsite 3D Printing Technology (CONPrint3D) / Venkatesh Naidu Nerella and Viktor Mechtcherine - Construction 3D Printing Camille Holt, Laurie Edwards, Louise Keyte, Farzad Moghaddam and Belinda Townsend - Properties of Extrusion-Based 3D Printable Geopolymers for Digital Construction Applications / Behzad Nematollahi, Ming Xia, Praful Vijay and Jay G. Sanjayan - Industrial Adoption of 3D Concrete Printing in the Australian Market: Potentials and Challenges / Daniel Avrutis, Ali Nazari and Jay G. Sanjayan.
LC Subjects	Three-dimensional printing--Industrial applications. Concrete.
Notes	Includes bibliographical references and index.

Advances in ceramic matrix composites

LCCN	2018304138
Type of material	Book
Main title	Advances in ceramic matrix composites / edited by I.M. Low.
Edition	Second edition.
Published/Produced	Duxford, United Kingdom: Woodhead Publishing, an imprint of Elsevier, [2018] ©2018
Description	xvii, 824 pages: illustrations (some color); 23 cm.
ISBN	0081021666 9780081021668
LC classification	TA418.9.C6 A286 2018
Related names	Low, It-Meng, editor.
Contents	1. Advances in ceramic matrix composites: introduction / I.M. Low - 1.1. The importance of ceramic matrix composites - 1.2. Novel material systems - 1.3. Innovative processing techniques -

1.4. Emerging and potential applications - References - 2. Processing, properties and applications of ceramic matrix composites, SiCf/SiC: an overview / Donghai Ding - 2.1. Introduction - 2.2. Novel interphase materials and new fabrication methods for traditional interphase materials - 2.3. Novel matrix manufacturing processes - 2.4. Nano-reinforcement - 2.5. Dielectric properties and microwave-absorbing applications - 2.6. Conclusion and future trends - References - 3. Nanoceramic matrix composites: types, processing, and applications / Jagannath Roy - 3.1. Introduction - 3.2. Nanostructured composite materials - 3.3. Bulk ceramic nanocomposites - 3.4. Nanoceramic composite coatings - 3.5. Conclusions - References - 4. Al2O3-SiC nanocomposites: preparation, microstructure, and properties / Dusan Galusek - 4.1. Introduction - 4.2. Experimental methods - 4.3. Results and discussion - 4.4. Conclusions - Acknowledgments - References - 5. Advances in manufacture of ceramic matrix composites by infiltration techniques / Dmitri Kopeliovich - 5.1. Introduction - 5.2. Classification of infiltration techniques - 5.3. Reinforcing fibers - 5.4. Interphases - 5.5. Polymer infiltration and pyrolysis - 5.6. Chemical vapor infiltration - 5.7. Reactive melt infiltration - 5.8. Slurry Infiltration - 5.9. Sol-gel infiltration - 5.10. Combined infiltration methods - 5.11. Future trends in fabrication of ceramic matrix composites by infiltration methods - References - Further reading - 6. Manufacture of graded ceramic matrix composites by infiltration technique / I.M. Low - 6.1. Introduction - 6.2. Processing and characterization techniques - 6.3. Phase composition profiles and mechanical properties - 6.4. Conclusions - Acknowledgments - References - Further reading - 7. Heat treatment for strengthening silicon carbide ceramic matrix

natural fiber - reinforced cement composites / I.M. Low - 12.1. Introduction - 12.2. Experimental procedure - 12.3. Results and discussion - 12.4. Conclusions - References - 13. Recent progress in development of high-performance tungsten carbide-based composites: synthesis, characterization, and potential applications / Suyan Li - 13.1. Introduction - 13.2. Present research work on WC-based composites - 13.3. Strategies to enhance the performance of WC-based composites - 13.4. Typical applications of WC-based composites - 13.5. Conclusions and outlook - References - 14. Double-A layered ceramics / Chunfeng Hu - 14.1. Introduction - 14.2. Experimental procedures - 14.3. Results and discussion - 14.4. Conclusions - References - 15. Role of interfaces in mechanical properties of ceramic matrix composites / Zhenhai Xia - 15.1. Introduction - 15.2. Interfaces in ceramic matrix composites - 15.3. Toughening and strengthening mechanisms in ceramic matrix composites - 15.4. Engineering design of interfaces for high strength and toughness - 15.5. Concluding remarks - Acknowledgments - References - 16. Using finite element analysis to understand the mechanical properties of ceramic matrix composites / Besim Ben-Nissan - 16.1. Introduction - 16.2. The use of finite element analysis to study ceramic matrix composites - 16.3. Concluding remarks - References - 17. Understanding the wear and tribological properties of ceramic matrix composites / Chaohui Zhang - 17.1. Introduction - 17.2. Friction - 17.3. Lubrication - 17.4. Wear - 17.5. Friction and wear of ceramics - 17.6. Tribological properties of ceramic matrix composites (CMCs) - 17.7. Future trends - Sources of further information and advice - References - 18. Understanding and improving the thermal stability of layered ternary carbides and nitrides / I.M. Low - 18.1. Introduction - 18.2.

High-temperature stability of Ti3SiC2 - 18.3. High-temperature stability of Ti3AlC2 and Ti2AlC - 18.4. High-temperature stability of ternary nitrides - 18.5. Testing the thermal stability of layered ternary carbides - 18.6. The high-temperature stability of particular layered ternary carbides - 18.7. Conclusions - 18.8. Future trends - Acknowledgments - References - Further reading - 19. Advances in geopolymer composites with natural reinforcement / I.M. Low - 19.1. Introduction - 19.2. Experimental procedures - 19.3. Results and discussion - 19.4. Conclusions - Acknowledgments - References - 20. Advances in self-healing ceramic matrix composites / F. Rebillat - 20.1. Introduction - 20.2. Understanding oxidation behaviour - 20.3. Understanding self-healing - 20.4. Issues in processing self-healing ceramic matrix composites - 20.5. The design of the interphase and matrix architectures - 20.6. Assessing the properties of self-healing ceramic matrix composites - 20.7. Testing the oxidation of self-healing matrix composites - 20.8. Self-healing silicate coatings - 20.9. Modelling self-healing - 20.10. Applications - 20.11. Trends in the development of self-healing composite materials - 20.12. Conclusion - References - 21. Self-crack-healing behavior in ceramic matrix composites / Kotoji Ando - 21.1. Introduction - 21.2. Material design for self-crack-healing - 21.3. Influence of oxygen partial pressure on self-crack-healing - 21.4. Influence of oxygen partial pressure on self-crack-healing under stress - 21.5. Conclusion - References - 22. Geopolymer (aluminosilicate) composites: synthesis, properties and applications / M. Welter - 22.1. Introduction - 22.2. Geopolymer matrix composite materials - 22.3. Processing geopolymer composites - 22.4. Properties of geopolymers and geopolymer composites - 22.5. Applications - 22.6. Future trends - References -

23. Fibre-reinforced geopolymer composites (FRGCs) for structural applications / Faiz Uddin Ahmed Shaikh - 23.1. Introduction - 23.2. Source materials used for geopolymers - 23.3. Alkaline solutions used for geopolymers - 23.4. Manufacturing FRGCs - 23.5. Mechanical properties of FRGCs - 23.6. Durability of FRGCs - 23.7. Future trends - 23.8. Conclusion - References - 24. Ceramic matrix composites in fission and fusion energy applications / Yutai Katoh - 24.1. Introduction - 24.2. Radiation effects on ceramic matrix composite - 24.3. Small specimen test technology and constitutive modeling - 24.4. Fusion energy - 24.5. Fission energy - 24.6. Future prospects - 24.7. Further sources of information - References - 25. The use of ceramic matrix composites for metal cutting applications / Jun Zhao - 25.1. Introduction - 25.2. Classification of ceramic matrix composites (CMCs) for metal cutting applications - 25.3. Strengthening and toughening of ceramic tool materials - 25.4. Design and fabrication of graded ceramic tools - 25.5. Application of ceramic inserts in the machining of hard-to-cut materials - 25.6. Future trends - Acknowledgments - References - 26. Cubic boron nitride-containing ceramic matrix composites for cutting tools / Takashi Goto - 26.1. Introduction - 26.2. Densification and relative density - 26.3. Microstructures - 26.4. Mechanical properties - 26.5. Phase transformation of cBN to hBN - 26.6. Conclusion and future trends - References 27. Multilayer glass - ceramic composites for microelectronics: processing and properties / Rosidah Alias - 27.1. Introduction - 27.2. Testing multilayer glass - ceramic composites - 27.3. Key challenges in preparing multilayer glass - ceramic composites - 27.4. Evaluation of fabricated glass - ceramic substrates - 27.5. Conclusion - Acknowledgments - References - 28.

Microfabrication of components based on functionally graded materials / Kyle Jiang - 28.1. Introduction - 28.2. Experimental - 28.3. Results and discussion - 28.4. Conclusions - References - 29. Ceramics in restorative dentistry / Richard Stoll - 29.1. Introduction - 29.2. Development of ceramics for restorative dentistry - 29.3. Dental bioceramics - 29.4. Dental CAD/CAM systems - 29.5. Clinical adjustments - 29.6. Surface integrity and reliability of ceramic restorations - 29.7. Conclusion - Acknowledgments - References - 30. Resin-based ceramic matrix composite materials in dentistry / Hesam Mirmohammadi - 30.1. Introduction - 30.2. The development of dental composites - 30.3. Composition of dental composites - 30.4. Classification of dental composites - 30.5. Limitations of dental composites - 30.6. The development of nanocomposites - 30.7. Indirect dental composites - 30.8. Resin-based composite cements - 30.9. Environmental factors influencing dental composites - 30.10. Future trends - References - 31. Ceramics in solid oxide fuel cells for energy generation / Xu Xu - 31.1. Solid oxide fuel cells - 31.2. Electrolytes - 31.3. Anodes of solid oxide fuel cells - 31.4. Cathode materials of solid oxide fuel cells - 31.5. Interconnects of solid oxide fuel cell - 31.6. Sealants of solid oxide fuel cell - 31.7. Conclusions - References.

LC Subjects	Ceramic-matrix composites.
Other Subjects	Ceramic-matrix composites.
Notes	Includes bibliographical references and index.
Series	Woodhead publishing series in composites science and engineering Woodhead Publishing series in composites science and engineering.

An introduction to aluminosilicates

LCCN	2020000118
Type of material	Book
Main title	An introduction to aluminosilicates / Nero Regina Blevins, editor.
Published/Produced	New York: Nova Science Publishers, [2020]
Description	xvi, 480 pages; 27 cm
ISBN	9781536172508 (hardcover) (adobe pdf)
LC classification	QD181.S6 I56 2020
Related names	Blevins, Nero Regina, editor.
Summary	"An Introduction to Aluminosilicates first reviews the main advances in the methods of obtaining synthetic montmorillonite for environmental and biomedical applications. Polyhydroxy metal complexes of Al, Fe, Zr, Cr and Ti (inorganic pillarizing agents) are studied for the synthesis of pillared clays. Information is provided on the possibility of using amorphous aluminosilicates as a modifying additive in the formulation of cement-based dry building mixtures. The physicochemical properties of the additive are considered, and the results of XRD and DTA analyzes are presented. The authors investigate the molecular orientation and surface morphology of organized molecular films with regard to solid-state structures for organo-modified aluminosilicates by a surface pressure-area isotherm, in-plane and out-of plane X-ray diffraction, and atomic force microscopy. A new method for the synthesis of amorphous mesoporous aluminosilicates with acidic properties and a narrow pore size distribution in the range of 2-7 nm is also explored. The urease immobilization of composite adsorbents polyacrylamide-bentonite, polyacrylamide-chitosan, and polyacrylamide-chitosan-bentonite is assessed, produced from polyacrylamide as a hydrogel, bentonite as an aluminosilicate mineral, and chitosan as a polysaccharide. Following this, the authors provide

a brief overview of literature data where crystalline aluminosilicates have been utilized for the capture and immobilization of radionuclides. The physiochemical properties of as-synthesized layered aluminosilicate are characterized by small angle PXRD, FTIR, porosity studies, thermal studies, 27Al MAS NMR, and morphological studies. Subsequently, this compilation addresses the way in which magnetic nanoparticles have proven to be frequently occurring building blocks for hybrid structure construction and, moreover, have shown innovative prospects as multifunctional adsorbents. Modern methods for the synthesis of highly dispersed micro- and micromesoporous mordenite with a high degree of crystallinity, as well as high crystallinity mordenite with a hierarchical porous structure, methods for modifying the mentioned crystalline aluminosilicates as a result of postsynthetic treatments, and modern adsorption and catalytic systems based on them are described. Next, the high activity and selectivity of high crystallinity zeolite Y with a hierarchical porous structure in the synthesis of practically important oligomers of various unsaturated compounds is established. The adsorptive features of PAAm-Ch-Z for Th4 are investigated in view of its dependency on pH, concentration, time, temperature and ionic strength. The parameters derived from the compatibility of experimentally obtained data to Langmuir, Freundlich and Dubinin-Radushkevich, van t'Hoff, pseudo frist-/second-order and Weber Morris models are utilized in the evaluation of adsorption and its thermodynamics and kinetics. Based on the importance and wide use of industrial kaolin, the penultimate chapter focuses on two materials that may be produced from metakaolin: geopolymers and synthetic zeolites. The applications of kaolin in the production of geopolymers has been studied in order to obtain geopolymer matrices with different application potentials. Lastly, the authors consider a possible mechanism for the generation of an acetyl-zeolite intermediate via transfer from different acetyl donors through theoretical studies using a cluster model of H-ZSM-5 zeolite designed by three TO4 tetrahedral units"-- Provided by publisher.

Contents Montmorillonite-based aluminosilicate materials: properties and applications / Domingos Lusitâneo Pier Macuvele, Ana Paula Fagundes, Rozineide Aparecida Antunes Boca Santa, Jonas Valente Matsinhe, Natan Padoin, Cíntia Soares, and Humberto Gracher Riella - Pillared Clays, a class of heterogeneous catalysts derived from modified clay minerals: properties and uses in waste pyrolysis / Ana C. S. Serra, Monica R. C. Marques and Jacyra G. Faillace - Application of synthetic aluminosilicates in adhesive dry construction mixtures / Loganina Valentina Ivanovna and Zhegera Kristina Vladimirovna - Surface modification of aluminosilicate by amphiphiles: fabrication of ultrathin films / Yoshinori Abiko and Atsuhiro Fujimori - Amorphous mesoporous aluminosilicates in the synthesis of N-Heterocyclic compounds / Nellya G. Grigorieva, Marat R. Agliullin, Sergey V. Bubennov, Svetlana A. Kostyleva, Nadezhda A. Filippova, Vera R.Bikbaeva and Boris I. Kutepov - Composites of bentonite, chitosan and polyacrylamide: preparation, characterization, and application in urease immobilization / Demet Baybaş, Serhat Öner and Ulvi Ulusoy - Crystalline aluminosilicates for radionuclide remediation / Brian J. Riley and Saehwa Chong - Synthesis of layered aluminosilicate by direct method / T.T. Sundaram - Clinoptilolite-rich tuff - a native aluminosilicate nanocomposite for the removal of trace quantities of pollutants / Eva Chmielewská,

	Renáta Górová, Włodzimierz Tylus, Mária Kovaľaková and Jozef Kravčák - Modern syntheses and the use of MOR type zeolite in adsorption and catalysis / Travkina Olga, Ahmed K. Ramadan and Kutepov Boris - Oligomerization of unsaturated compounds in the presence of hierarchical zeolite Y / Sergey V. Bubennov, Nellya G. Grigoriev, Dmitry V. Serebrennikov, Marat R. Agliullin, Alfira N. Khazipova and Boris I. Kutepov - Preparation and characterization of chitosan-zeolite composite@polyacrylamide: adsorptive features for thorium / Onur Eninanç and Ulvi Ulusoy - Application of kaolin as an aluminosilicate source to produce geopolymer and zeolitic materials / Rozineide Aparecida Antunes Boca Santa PhD, Humberto Gracher Riella, Domingos Lusitâneo Pier Macuvele, Natan Padoin, and Cíntia Soares - A DFT study of the acetyl transfer reaction catalyzed by H-ZSM-5 Zeolite using a 3T-cluster approach / Pablo Corregidor and Emilce Ottavianelli.
LC Subjects	Aluminum silicates.
Notes	Includes bibliographical references and index.
Additional formats	Online version: Introduction to aluminosilicates New York: Nova Science Publishers, Inc., [2019] 9781536173109 (DLC) 2020000119
Series	Geology and mineralogy research developments

An introduction to aluminosilicates

LCCN	2020000119
Type of material	Book
Main title	An introduction to aluminosilicates / [edited by Nero Regina Blevins].
Published/Produced	New York: Nova Science Publishers, Inc., [2019]
Description	1 online resource
ISBN	9781536173109 (adobe pdf) (hardcover)
LC classification	QD181.S6
Related names	Blevins, Nero Regina, editor.

Summary

"An Introduction to Aluminosilicates first reviews the main advances in the methods of obtaining synthetic montmorillonite for environmental and biomedical applications. Polyhydroxy metal complexes of Al, Fe, Zr, Cr and Ti (inorganic pillarizing agents) are studied for the synthesis of pillared clays. Information is provided on the possibility of using amorphous aluminosilicates as a modifying additive in the formulation of cement-based dry building mixtures. The physicochemical properties of the additive are considered, and the results of XRD and DTA analyzes are presented. The authors investigate the molecular orientation and surface morphology of organized molecular films with regard to solid-state structures for organo-modified aluminosilicates by a surface pressure-area isotherm, in-plane and out-of plane X-ray diffraction, and atomic force microscopy. A new method for the synthesis of amorphous mesoporous aluminosilicates with acidic properties and a narrow pore size distribution in the range of 2-7 nm is also explored. The urease immobilization of composite adsorbents polyacrylamide-bentonite, polyacrylamide-chitosan, and polyacrylamide-chitosan-bentonite is assessed, produced from polyacrylamide as a hydrogel, bentonite as an aluminosilicate mineral, and chitosan as a polysaccharide. Following this, the authors provide a brief overview of literature data where crystalline aluminosilicates have been utilized for the capture and immobilization of radionuclides. The physiochemical properties of as-synthesized layered aluminosilicate are characterized by small angle PXRD, FTIR, porosity studies, thermal studies, 27Al MAS NMR, and morphological studies. Subsequently, this compilation addresses the way in which magnetic nanoparticles have proven to be frequently occurring building blocks for hybrid structure construction and, moreover, have shown innovative prospects as multifunctional adsorbents. Modern methods for the synthesis of highly dispersed micro- and micromesoporous mordenite with a high degree of crystallinity, as well as high crystallinity mordenite with a hierarchical porous structure, methods for modifying the mentioned crystalline aluminosilicates as a result of postsynthetic treatments, and modern adsorption and catalytic systems based on them are described. Next, the high activity and selectivity of high crystallinity zeolite Y with a hierarchical porous structure in the synthesis of practically important oligomers of various unsaturated compounds is established. The adsorptive features of PAAm-Ch-Z for Th4 are investigated in view of its dependency on pH, concentration, time, temperature and ionic strength. The parameters derived from the compatibility of experimentally obtained data to Langmuir, Freundlich and Dubinin-Radushkevich, van t'Hoff, pseudo frist-/second-order and Weber Morris models are utilized in the evaluation of adsorption and its thermodynamics and kinetics. Based on the importance and wide use of industrial kaolin, the penultimate chapter focuses on two materials that may be produced from metakaolin: geopolymers and synthetic zeolites. The applications of kaolin in the production of geopolymers has been studied in order to obtain geopolymer matrices with different application potentials. Lastly, the authors consider a possible mechanism for the generation of an acetyl-zeolite intermediate via transfer from different acetyl donors through theoretical studies using a cluster model of H-ZSM-5 zeolite designed by three TO4 tetrahedral units"-- Provided by publisher.

Contents Montmorillonite-based aluminosilicate materials: properties and applications / Domingos Lusitâneo Pier Macuvele, Ana Paula Fagundes, Rozineide

Aparecida Antunes Boca Santa, Jonas Valente Matsinhe, Natan Padoin, Cíntia Soares, and Humberto Gracher Riella - Pillared Clays, a class of heterogeneous catalysts derived from modified clay minerals: properties and uses in waste pyrolysis / Ana C. S. Serra, Monica R. C. Marques and Jacyra G. Faillace - Application of synthetic aluminosilicates in adhesive dry construction mixtures / Loganina Valentina Ivanovna and Zhegera Kristina Vladimirovna - Surface modification of aluminosilicate by amphiphiles: fabrication of ultrathin films / Yoshinori Abiko and Atsuhiro Fujimori - Amorphous mesoporous aluminosilicates in the synthesis of N-Heterocyclic compounds / Nellya G. Grigorieva, Marat R. Agliullin, Sergey V. Bubennov, Svetlana A. Kostyleva, Nadezhda A. Filippova, Vera R.Bikbaeva and Boris I. Kutepov - Composites of bentonite, chitosan and polyacrylamide: preparation, characterization, and application in urease immobilization / Demet Baybaş, Serhat Öner and Ulvi Ulusoy - Crystalline aluminosilicates for radionuclide remediation / Brian J. Riley and Saehwa Chong - Synthesis of layered aluminosilicate by direct method / T.T. Sundaram - Clinoptilolite-rich tuff - a native aluminosilicate nanocomposite for the removal of trace quantities of pollutants / Eva Chmielewská, Renáta Górová, Włodzimierz Tylus, Mária Kovaľaková and Jozef Kravčák - Modern syntheses and the use of MOR type zeolite in adsorption and catalysis / Travkina Olga, Ahmed K. Ramadan and Kutepov Boris - Oligomerization of unsaturated compounds in the presence of hierarchical zeolite Y / Sergey V. Bubennov, Nellya G. Grigoriev, Dmitry V. Serebrennikov, Marat R. Agliullin, Alfira N. Khazipova and Boris I. Kutepov - Preparation and characterization of chitosan-zeolite composite@polyacrylamide:

	adsorptive features for thorium / Onur Eninanç and Ulvi Ulusoy - Application of kaolin as an aluminosilicate source to produce geopolymer and zeolitic materials / Rozineide Aparecida Antunes Boca Santa PhD, Humberto Gracher Riella, Domingos Lusitâneo Pier Macuvele, Natan Padoin, and Cíntia Soares - A DFT study of the acetyl transfer reaction catalyzed by H-ZSM-5 Zeolite using a 3T-cluster approach / Pablo Corregidor and Emilce Ottavianelli.
LC Subjects	Aluminum silicates.
Notes	Includes bibliographical references and index.
Additional formats	Print version: An introduction to aluminosilicates New York: Nova Science Publishers, Inc., [2019] 9781536172508 (DLC) 2020000118
Series	Geology and mineralogy research developments

Calcined Clays for Sustainable Concrete: Proceedings of the 2nd International Conference on Calcined Clays for Sustainable Concrete

LCCN	2019769366
Type of material	Book
Main title	Calcined Clays for Sustainable Concrete: Proceedings of the 2nd International Conference on Calcined Clays for Sustainable Concrete / edited by Fernando Martirena, Aurélie Favier, Karen Scrivener.
Edition	1st ed. 2018.
Published/Produced	Dordrecht: Springer Netherlands: Imprint: Springer, 2018.
Description	1 online resource (XXVI, 520 pages 306 illustrations) text file PDF
ISBN	9789402412079
Related names	Favier, Aurélie, editor. Martirena, Fernando, editor. Scrivener, Karen, editor.
Summary	This volume focuses on research and practical issues linked to Calcined Clays for Sustainable Concrete. The main topics are geology of clays,

hydration and performance of blended system with calcined clays, alkali activated binders, applications in concrete and mortar, durability of concrete under various aggressive conditions, and economic and environmental impacts of the use of calcined clays in cement based materials. This book compiles the different contributions of the 2nd International Conference on Calcined Clays for Sustainable Concrete, which took place in La Habana, December 5th-7th, 2017.The papers update the latest research in their field, carried out since the last conference in 2015. Overall it gives a broad view of research on calcined clays and their application in the field of construction, which will stimulate further research into calcined clays for sustainable concrete.

Contents Foreword - Calcined Clay-Cement Stabilisation - Physicochemical Attributes and Stabilised Strengths of A-1-a and A-2-6 Soils, by O. A. Adekitan and G. M. Ayininuola - Sulfate Resistance of Cement Mortar Containing Metakaolin by Abdelsalam M. Akasha and Jamal M. Abdullah - Use of R3 Rapid Screening Test to Determine Reactivity and Chloride Binding Potential of Locally Available Kaolinite Clay, by M. Almarshoud, J.L. Saint Rome, K. A. Riding - Assessment of cuban kaolinitic clays as source of supplementary cementitious materials to production of cement based on clinker - calcined clay - limestone, by Roger S. Almenares Reyes, Adrián Alujas Díaz, Sergio Betancourt Rodríguez, Carlos Alberto Leyva Rodríguez, José Fernando Martirena Hernández - Proposal of a methodology for the preliminary assessment of kaolinitic clay deposits as a source of SCMs, by Adrián Alujas Díaz, Roger S. Almenares Reyes, Florencio Arcial Carratalá, José F. Martirena Hernández - Hydration study of Limestone Calcined Clay Cement (LC3) using various grades of calcined kaolinitic clays, by

F. Avet, K. Scrivener - Reaction degree of metakaolin in Limestone Calcined Clay Cement (LC3), by F. Avet, K. Scrivener - Durability of steam cured pozzolanic mortars at atmospheric pressure, by Kübra Ekiz Barış and Leyla Tanaçan - Use of Low-Carbon Cement in the Preparation of Masonry Mortars for Building Restoration, by D. Betancourt Cura, Y. Lima Triana, F. Martirena Hernandez - Quantification of Pore Size Distribution Modification due to Metakaolin Inclusion in Cement Based Systems, by B Bhattacharjee - Limestone Calcined Clay Cement: The experience in India this far, by Shashank Bishno1 and Soumen Maity - Pilot scale production of Limestone Calcined Clay Cement, by Shashank Bishnoi, Soumen Maity, Mukesh Kumar, SK Saxena, SK Wali - The special case of North-Eastern India for the production of LC3, by Shashank Bishnoi, Soumen Maity, S. P. Pandey, P. K. Tripathy - Comparative study of compressive creep behavior of concrete with metakaolin or silica fume, by R. Bucher, H. Cagnon, T. Vidal, M. Cyr - Effect of carbonate minerals and calcination of carbonatites and kamafugites on their pozzolanic performance and early age concrete properties, by A. Buregyeya, Y. Ballim, S. Nwaubani, A. G. Kerali, M. Otieno - Assessment of the Pozzolanic Reactivity of Calcined Kaolinitic Clays by a Rapid Alkaline Solubility Test, by E. Cabrera, R. Almenares, A. Alujas - Sustainability of Cuban Construction Supply Chain by means of LC3 Cement: Case Studies in Villa Clara Province, by Yudiesky Cancio Díaz, Inocencio Raúl Sánchez Machado, José Fernando Martirena Hernández, Guillaume Habert - Degradation of Calcined Clay-Limestone Cementitious Composites under Sulfate Attack, by YU Cheng, YUAN Peng, YU Xin, LIU Jiaping - Sulfate and alkali-silica performance of blended cements containing illitic calcined clays, by Gisela Cordoba, Agustin Rossetti, Dario Falcone, E.F: Irassar - Use of Ceramic Waste as a Pozzolanic Addition on Cement, by Ing. Rayda Crespo Castillo - Development of the microstructure in LC3 systems and its effect on concrete properties, by Yuvaraj Dhandapani1, Vignesh K., Thangadurai Raja and Manu Santhanam - Carbonation of Concrete with Low Carbon Cement LC3 Exposed to Different Environmental Conditions, by Ernesto Díaz, Raúl González, Dayran Rocha Adrian Alujas, Fernando Martirena - Evaluation of compressive strength and microstructure of cement pastes containing different qualities of metakaolin, by N. Dumani and J. Mapiravana - Influence of initial water curing on strength and microstructure development of blended cements, by A. C. Emmanuel, H. Talluru, S. Krishnan and S. Bishnoi - Initial performance evaluation of calcined clay based ternary blended cement under various climatic conditions in India, by A. C. Emmanuel, G. Mishra, S. Bishnoi - Alkali Silica Reaction and Sulfate Attack: Expansion of Limestone Calcined Clay Cement, by A. Favier and K. Scrivener - The effect of limestone on the performance of ternary blended cement LC3: limestone, calcined clays and cement, by Aurélie Favier, Franco Zunino, Ioannis Katrantzis and Karen Scrivener - Influence of Clay Type on Performance of Calcined Clay - Limestone Portland Cements, by S. Ferreiro, D. Herfort and J. S. Damtoft - Metakaolin-based geopolymers for nuclear waste encapsulation, by D. A. Geddes, X. Ke, S. A. Bernal, M. Hayes and J. L. Provis - The influence of recycled concrete and clay brick particles on the strength and porosity of cement-based pastes, by T.M. Grabois, G.C. Cordeiro and R.D. Toledo Filho - The Effect of Kaolinite Content of China Clay on the Reactivity of Limestone Calcined Clay Cement, by P. K. Haldar,

S. Mithia, K. Mukherjee, N. R. Dhabarde, E. Bansal, P. Phulwari, A. Kumar, S. Kesh, S. Maity - The effect of alkali on the properties of Limestone Calcined Clay Cement (LC3), by W. Hanpongpun and K. Scrivener - On the Reactivity of Calcined Clays from Lower Lusatia for the Production of Durable Concrete Structures, by Klaus-Juergen Huenger, Robert Gerasch, Ingolf Sander, Maria Brigzinsky - Compressive strength improvements of cement-based composites achieved with additional milling of metakaolin, by Biljana Ilić, Aleksandra Mitrović, Vlastimir Radonjanin, Mirjana Malešev, Miodrag Zdujić - Properties of the cement-based composites with high content of metakaolin, by Biljana Ilić, Vlastimir Radonjanin, Mirjana Malešev, Miodrag Zdujić, Aleksandra Mitrović - Calcined clays - performance as composite material, by Christian Kalb - Structural ordering of aged and hydrothermally cured metakaolin based potassium geopolymers, by Xinyuan Ke, John L. Provis and Susan A. Bernal - Carbonation of limestone calcined clay cement concrete, by M.S.H. Khan, Q.D. Nguyen, A. Castel - Grinding of calcined clays and its effects on cement properties, by W. Kluge, Dr. B. O. Assmann - Hydration and mechanical properties of limestone calcined clay cement produced with marble dust, by Sreejith Krishnan, Arun C Emmanuel, Swadesh Kumar Kanaujia and Shashank Bishnoi - Performance-based Design Procedure Applied to the Selection of Low-CO2 Binder Systems Including Calcined Clay, by Wilson R. Leal da Silva, Lars N. Thrane, Thomas L. Svensson, Sergio Ferreiro, Duncan Herfort, Claus Padel, Jesper S. Damtoft - Thermal Processing of Calcined Clay, by J. Lemke, C. Berger - Thermal transformation of illitic-chlorite clay and its pozzolanic activity, by Roxana Lemma, Cristina C. Castellano, Viviana L. Bonavetti,

Monica A. Trezza, Viviana F. Rahhal and Edgardo F. Irassar - In-situ observation of dissolution behavior of carbonatite in water glass solution, by Jing Li, Jianqin Lin, Qijun Yu, Jie Hu, Suhong Yin - Analysis of the Mixing Performance Containing the LC3 as Agglomerant with Different Types of Calcined Clay, by D. Lins, J. Rêgo and E. Silva - Evaluation of calcined clays from Boyaca-Colombia containing alunite as Supplementary Cementitious Materials, by Ariam Lozano Perez, Mathieu Antoni - Improvement of the environmental energy sustainability in the production of cement Portland with addition of thermally activated clays, by Dr. Eng. I. L. Machado, Dr. Eng. H. I. Moya, Lic. S. B. Sánchez, Dr. Sc. Eng. F. Martirena - Resource mapping of china clay for LC3 application in India, by Soumen Maity and Santanu Mithia - Chloride Transport Behavior of LC3 Binders, by H. Maraghechi, F. Avet, and K. Scrivener - Blended cements with calcined illitic clay: workability and hydration, by Guillermina Marchetti, Jaroslav Pokorny, Alejandra Tironi, Mónica A. Trezza, Viviana F. Rahhal, Zbyšek Pavlík, Robert Černý, Edgardo F. Irassar - Low Carbon Cement LC3 in Cuba: ways to achieve a sustainable growth of cement production in emerging economies, by Fernando Martirena, Karen Scrivener - Studies on the influence of limestone-calcined clay blend on the hydration of cement, by G. Mishra, A. Emmanuel, S. Bishnoi - Thermal Resistivity of Chemically Activated Calcined Clays -Based Cements, by Marangu J. Mwiti, Thiong'o J. Karanja, Wachira J. Muthengia - Promising Early age evaluations of Fly ash - Calcined marl - OPC ternary cement, by Serina Ng and Tone Østnor - Applicability of lime reactivity strength potential test for the reactivity study of limestone calcined clay cement, by Anuj Parashar, Vineet Shah, Shashank Bishnoi -

Limestone and Calcined Clay Blended Cement Used as Low-Cost Binder to Reduce Heat Production and Potential for Delayed Ettringite Formation, by Jerry M. Paris and Christopher C. Ferraro - Hydrate Phase Assemblages in Calcium Sulfoaluminate - Metakaolin - Limestone Blends, by M. Pedersen, B. Lothenbach, F. Winnefeld, and J. Skibsted - Influence grinding procedure, limestone content and PSD of components on properties of clinker-calcined clay-limestone cements produced by intergrinding, by A. Pérez, A. Favier, K. Scrivener and F. Martirena - Durability of concretes made with calcined clay composite cements, by Roland Pierkes, Simone E. Schulze, Jörg Rickert - Alkali-activation of calcined clays - past, present and future, by John L. Provis - Influence of a calcined clay and the temperature on the hydration of an oilwell cement, by Juan Alberto Ribalta, Adrián Alujas Díaz and José Fernando Martirena - Influence of the kind of mineral addition and the seawater on the hydration of a Portland cement, by Juan Alberto Ribalta*, Leidys Laura Pérez and Adrián Alujas Díaz - Standardization strategy of Low Carbon Cement in Cuba. Case study for "Siguaney" Cement Factory, by D. Rocha, R. Almenares, S. Sanchez, A. Alujas, F. Martirena - Antibacterial metakaolin-based geopolymer cement, by Jose-Carlos Rubio-Avalos - Identification of reactive sites in calcined kaolinite and montmorillonite from a combination of chemical methods and solid-state NMR Spectroscopy, by C. Ruiz-Santaquiteria and J. Skibsted - The Decay of the Historical Site of Malecon in Havana, Cuba: Salt Crystallization Damage at Repair Interfaces, by A.M. Aguilar Sanchez, F. Caruso, F. Girardet, F. Martirena, T. Wangler, and R.J. Flatt

LC Subjects Building materials.
Economic development--Environmental aspects.

	Inorganic chemistry.
	Mineral resources.
Other Subjects	Building Materials.
	Development and Sustainability.
	Inorganic Chemistry.
	Mineral Resources.
	Structural Materials.
Additional formats	Print version: Calcined clays for sustainable concrete. 9789402412062 (DLC) 2017954894
	Printed edition: 9789402412062
	Printed edition: 9789402412086
	Printed edition: 9789402415094
Series	RILEM Bookseries, 2211-0844; 16
	RILEM Bookseries, 2211-0844; 16

Clay nanoparticles: properties and applications

LCCN	2021444780
Type of material	Book
Main title	Clay nanoparticles: properties and applications / edited by Giuseppe Cavallaro, Department of Physics and Chemistry, University of Palermo, Palermo, Italy, Rawil Fakhrullin, Institute of Fundamental Medicine and Biology, Kazan Federal University, Kazan, Republic of Tatarstan, Russian Federation, Pooria Pasbakhsh, Disipline of Mechanical Engineering, School of Engineering, Monash University, Selangor, Malaysia.
Published/Produced	Amsterdam, Netherlands; Cambridge, MA: Elsevier, [2020]
Description	xv, 352 pages: illustrations; 24 cm
ISBN	9780128167830
	0128167831
Related names	Cavallaro, Giuseppe.
	Fakhrullin, Rawil.
	Pasbakhsh, Pooria.
Summary	Clay Nanoparticles: Properties and Applications sets out the major properties of clay nanoparticles and their technological applications. The first part of the book focuses on the characterization of

	nanoclays, including layered, fibrous and tubular clay minerals. The second part illustrates the current and potential applications of nanoclays within material science and biotechnology. These include the development of geopolymers and bionanocomposites based on sustainable polymers filled with ecocompatible nanoclay. The potential use of nanoclays as flame retardants is also discussed, along with the correlation between the properties and potential applications of several nanoclay types.
LC Subjects	Nanostructured materials. Clay.
Other Subjects	Clay Nanostructured materials
Notes	Includes bibliographical references and index.
Additional formats	ebook version: 9780128167847
Series	Micro and nano technologies series Micro & nano technologies.

Developments in strategic materials and computational design III: a collection of papers presented at the 36th International Conference on Advanced Ceramics and Composites, January 22-27, 2012, Daytona Beach, Florida

LCCN	2012554455
Type of material	Book
Meeting name	International Conference on Advanced Ceramics and Composites (36th: 2012: Daytona Beach, Fla.)
Main title	Developments in strategic materials and computational design III: a collection of papers presented at the 36th International Conference on Advanced Ceramics and Composites, January 22-27, 2012, Daytona Beach, Florida / edited by Waltraud M. Kriven ... [et al.]; volume editors Michael Halbig, Sanjay Mathur.
Published/Created	Hoboken, N.J.: John Wiley & Sons; Westerville, Ohio: American Ceramic Soc., c2013.
Description	xi, 202 p.: ill.; 24 cm.

Links	Publisher description http://www.loc.gov/catdir/enhancements/fy1313/2012554455-d.html Table of contents only http://www.loc.gov/catdir/enhancements/fy1403/2012554455-t.html
ISBN	9781118206003 1118206002
LC classification	TA455.C43 I559 2012 TP785 .A55 2012 vol. 33 no.10
Variant title	ICACC Developments in strategic materials and computational design three Developments in strategic materials and computational design 3
Related names	Kriven, Waltraud M. Halbig, Michael C., 1969- Mathur, Sanjay. American Ceramic Society.
LC Subjects	Ceramic materials--Congresses. Temperature control--Congresses. Detectors--Congresses.
Notes	Contains contributions from one symposium and three focused sessions that were part of the 36th International Conference on Advanced Ceramics and Composites held in Daytona Beach, Florida, January 2012. The session titles are: Thermal Management Materials and Technologies (Symposium 10); Geopolymers and other Inorganic Polymers; Computational Design, Modeling simulation and Characterization of Ceramic Composites; and, Advanced (Ceramic) Materials and Processing of Photonics and Energy. "American Ceramic Society" Includes bibliographical references and author index.
Series	Ceramic engineering and science proceedings, 0196-6219; v. 33, issue 10 Ceramic engineering and science proceedings; v. 33, no. 10.

Developments in strategic materials and computational design IV: a collection of papers presented at the 37th International Conference on Advanced Ceramics and Composites, January 27-February 1, 2013, Daytona Beach, Florida

LCCN	2016498536
Type of material	Book
Meeting name	International Conference on Advanced Ceramics and Composites (37th: 2013: Daytona Beach, Fla.)
Main title	Developments in strategic materials and computational design IV: a collection of papers presented at the 37th International Conference on Advanced Ceramics and Composites, January 27-February 1, 2013, Daytona Beach, Florida / edited by Waltraud M. Kriven, Jingyang Wang, Yanchun Zhou, Andrew L. Gyekenyesi; volume editors, Soshu Kirihara, Sujanto Widjaja.
Published/Produced	Hoboken, New Jersey: John Wiley & Sons, Inc., [2014] ©2014
Description	xiii, 322 pages: illustrations; 25 cm.
ISBN	9781118807279 1118807278
Variant title	Developments in strategic materials and computational design 4 Developments in strategic materials and computational design four
Related names	Kriven, Waltraud M., editor. Wang, Jingyang (Materials scientist), editor. Zhou, Yanchun, editor. Gyekenyesi, Andrew L., editor. Kirihara, Soshu, editor. Widjaja, Sujanto, editor.
LC Subjects	Ceramic materials--Congresses. Composite materials--Congresses. Strategic materials--Congresses.
Notes	"Contributions from ... Focused Session 1 - Geopolymers and Chemically Bonded Ceramics; Focused Session 2 - Thermal Management Materials and Technologies; Symposium 10 -

	Virtual Materials (Computational) Design and Ceramic Genome; and, Symposium 12 - Materials for Extreme Environments: Ultrahigh Temperature Ceramics and Nanolaminated Ternary Carbides and Nitrides."--Preface.
	"The American Ceramic Society."
	Includes bibliographical references and index.
Series	Ceramic engineering and science proceedings, 0196-6219; volume 34, issue 10
	Ceramic engineering and science proceedings; v. 34, no. 10.

Fire-Resistant Geopolymers: Role of Fibres and Fillers to Enhance Thermal Properties

LCCN	2019748311
Type of material	Book
Personal name	Vickers, Les. author.
Main title	Fire-Resistant Geopolymers: Role of Fibres and Fillers to Enhance Thermal Properties / by Les Vickers, Arie van Riessen, William D. A. Rickard.
Edition	1st ed. 2015.
Published/Produced	Singapore: Springer Singapore: Imprint: Springer, 2015.
Description	1 online resource (VIII, 129 pages 35 illustrations, 11 illustrations in color.)
	text file PDF
ISBN	9789812873118
Related names	Rickard, William D. A. author.
	van Riessen, Arie. author.
Summary	The book covers the topic of geopolymers, in particular it highlights the relationship between structural differences as a result of variations during the geopolymer synthesis and its physical and chemical properties. In particular, the book describes the optimization of the thermal properties of geopolymers by adding micro-structural modifiers such as fibres and/or fillers into the geopolymer matrix. The range of fibres and fillers used in geopolymers, their impact on the

microstructure and thermal properties is described in great detail. The book content will appeal to researchers, scientists, or engineers who are interested in geopolymer science and technology and its industrial applications.

Contents	Overview - History of Geopolymers - Portland Cement (OPC) and Concrete - Geopolymer Applications - Precursors and Additives for Geopolymer Synthesis - Geopolymer Chemistry - Fibres: Technical Benefits - Thermal Properties of Geopolymers - Fire Resistance of OPC and geopolymer - Conclusion.
LC Subjects	Building materials. Ceramics. Composite materials. Composites (Materials). Glass.
Other Subjects	Ceramics, Glass, Composites, Natural Materials. Building Materials. Structural Materials.
Additional formats	Print version: Fire-resistant geopolymers. 9789812873101 (DLC) 2014955620 Printed edition: 9789812873101 Printed edition: 9789812873125
Series	SpringerBriefs in Materials, 2192-1091 SpringerBriefs in Materials, 2192-1091

Fire-resistant geopolymers.

LCCN	2014955620
Type of material	Book
Main title	Fire-resistant geopolymers.
Published/Produced	New York, NY: Springer Berlin Heidelberg, 2015.
Links	Publisher description http://www.loc.gov/catdir/enhancements/fy1504/2014955620-d.html Table of contents only http://www.loc.gov/catdir/enhancements/fy1504/2014955620-t.html
ISBN	9789812873101

Geopolymer and Green Technology Special topic volume with invited peer reviewed papers only.

LCCN	2020488158
Type of material	Book
Main title	Geopolymer and Green Technology Special topic volume with invited peer reviewed papers only. edited by Mohd Mustafa Al Bakri Abdullah, Zarina Yahya and Liyana Jamaludin.
Published/Produced	Switzerland: Trans Tech Publications Ltd, 2016.
Description	236 p.: ill.
Links	https://doi.org/10.4028/www.scientific.net/KEM.673 Book description https://www.scientific.net/KEM.673/book
ISBN	9783038356417 3038356417
Related names	Mohd Mustafa Al Bakri Abdullah editor. Yahya, Zarina editor. Jamaludin, Liyana editor.
Summary	Collection of selected, peer reviewed papers from the special topic volume with invited peer reviewed papers only. The 23 papers are grouped as follows:Chapter 1: Using of Geopolymers and Pozzolanic MaterialsChapter 2: Polymers and CompositesChapter 3: Green Technologies in Chemical Engineering Keywords: Geopolymer, Alkali Activated, Green Technology, Waste Materials, Construction Industry Materials Science.
Notes	Includes bibliographical references and index.
Series	Key Engineering Materials 1013-9826 Vol. 673

Geopolymers and geocements: low environmentally impact ceramic materials: 12 international ceramics congress part H

LCCN	2012360160
Type of material	Book
Meeting name	International Meeting on Modern Ceramics Technologies (12th: 2010: Montecatini Terme, Italy)

Main title	Geopolymers and geocements: low environmentally impact ceramic materials: 12 international ceramics congress part H / edited by Pietro Vincenzini; co-edited by Cristina Leonelli.
Published/Created	Stafa-Zuerich; Enfield, NH: Trans Tech Pubs. ltd. on behalf of Techna Group, c2011.
Description	x, 202 p.: ill.; 24 cm.
ISBN	3908158524 9783908158523
LC classification	QD196 .I57 2010
Portion of title	Low environmentally impact ceramic materials
Related names	Vincenzini, P., 1939- Leonelli, Cristina. Forum on New Materials (5th: 2010: Montecatini Terme, Italy)
LC Subjects	Inorganic polymers--Congresses. Cement composites--Congresses. Ceramic materials--Congresses.
Notes	"Proceedings of the 12th International Ceramics Congress, part of CIMTEC 2010--12th International Ceramics Congress and 5th Forum on New Materials, Montecatini Terme, Italy, June 6-11, 2010." "Part H, including: Symposium CK--Geopolymers and geocements: low environmentally impact ceramic materials." Includes bibliographical references and indexes.
Series	Advances in science and technology, 1661-819X; vol. 69 Advances in science and technology; v. 69.

Ground Improvement Techniques and Geosynthetics: IGC 2016 Volume 2

LCCN	2019750935
Type of material	Book
Main title	Ground Improvement Techniques and Geosynthetics: IGC 2016 Volume 2 / edited by T Thyagaraj.
Edition	1st ed. 2019.

Published/Produced	Singapore: Springer Singapore: Imprint: Springer, 2019.
Description	1 online resource (XIII, 406 pages 249 illustrations, 62 illustrations in color.) text file PDF
ISBN	9789811305597
Related names	Thyagaraj, T. editor.
Summary	The book comprises select proceedings of the 2016 annual conference of the Indian Geotechnical Society (IGC 2016), with technical papers on the theme "Ground Improvement and Geosynthetics". The papers cover a wide range of topics, including chemical modification using admixtures, microbial-induced carbonate precipitation, geopolymers, fly ash and other industrial wastes, modification using geosynthetic materials such as natural and synthetic fibers, expanded polystyrene (EPS) geofoam, prefabricated vertical drains, geosynthetic encased-granular columns and mechanical densification through sand columns. This book is a valuable reference for researchers and practicing engineers alike.
Contents	THEME 3: GROUND IMPROVEMENT AND GEOSYNTHETICS - Behaviourial Study on Geopolymer Column in Soil - A Study of Parameters Influencing Efficiency of Micropile Groups - Alteration of CBR Values in Soft Soils using Enzymatic Lime - Factors which Influence PVD Performance in Soft Clay Consolidation - Experimental Study of Heave Control Technique for Expansive Soil using Micropiles and Geotextile Layers - Effect of Underground Void on the Internal Stress Distribution in Soil - Influence of Properties of Infill Material on the Behaviour of Geocells - Investigation of Glass Fiber Reinforcement Effect on the CBR Strength of Cohesive Soil - Bearing Capacity of Strip Footing on Clay Soil Reinforced with Metal Strips and with Anchors - Application of Microbial Induced

Carbonate Precipitation for Soil Improvement Via Ureolysis - Influence of Organic Content on Fly Ash Stabilisation of Clay - Strength Behaviours of the Clayey-Silt Soil Mixed with Fly Ash and Sand - Strength Behaviour Of Expansive Soil Treated With Quarry Dust And Ferric Chloride - Lime Stabilization of Subgrade with Waste Sand as Partial Soil Replacement - Assessment Of Vetiver Grass Root-Reinforcement In Strengthening The Soil - Geotechnical Aspects of Constructions Along the Canal Embankment using Rice Husk Ash as Stabilizer - Use of Jhama Columns as Replacement of Stone Columns - Laboratory Study on the Performance of Geo-Synthetic Reinforced Sand Bed - Repeated Load Response of Encapsulated Granular Trench Supported Footings on Embankments - Re-Appraisal of Physico-Mechanical Stability of Lime Treated Soils - Back analysis of PVD performance in Mumbai Port - Discharge and Absorption Capacity Tests on Composite Prefabricated Vertical Drains - Stabilization of Clay at Sunnam Cheruvu Area in Nadergul, Hyderabad using Organic Waste - The Tensile Strength Behavior of Lime Stabilized Soft Soil with Inclusion of Plastic Fiber - Pullout Behaviour of Geosynthetics - a Review of Laboratory Testing Techniques - Model Study on Cyclic Loading Responses of Flexible Pavement System Laid on Expansive Subgrade - Evaluating the Strength Characteristics of Lime and Metakaolin Stabilized Expansive Soil - Ground Improvement using Sand Columns to Mitigate Liquefaction - a Case Study - Stabilization of Expansive Soil with Red Mud and Lime - California Bearing Ratio and Permeability Behaviour of Fly Ash Reinforced with Geotextiles - Influence of Granulated Blast Furnace Slag Contents on California Bearing Ratio Value of Clay Gbfs Mixture - Geocell Reinforced Dense

Sand Bases Overlying Weak Sand Subgrades Under Repeated Loading - Load-Settlement Behaviour of Granular Pile on Cochin Marine Clay using Recycled Aggregate - Transmissivity of Coir Latex Composite - Influence of Combined Vertical and Horizontal Reinforcement on Granular Piles in Soft Clays - Effect of Ggbs and Lime on the Strength Characteristics of Black Cotton Soil - Soil Improvement using Microbial: a Review - Long Term Strength Studies on Natural Fibre Composite (N-F-C) Sheets for Use as Separator in Flexible Pavements in Terms of CBR Values - Coefficient of Consolidation for Vertical and Radial Drainage - Comparative Analysis of Strength Characteristics of Soil Reinforced with Coir and Polypropylene Fibers - Geo-Engineering Properties of Sedimented Flyash Bed Stabilized by Chemical Columns - A Short Review of Geosynthetic Granular Column Treatment of Soft Clay Soils - Engineered Anti Erosion Works Along the Right Bank of Jiabharali River in Assam - Static and Cyclic Properties of Eps Geofoam - Swell Behavior of Expansive Soils with Stabilized Fly Ash Columns.

LC Subjects Building.
Buildings-Design and construction.
Construction.
Engineering geology.
Engineering, Architectural.
Engineering--Geology.
Foundations.
Geotechnical engineering.
Hydraulics.

Other Subjects Geoengineering, Foundations, Hydraulics.
Building Construction and Design.
Geotechnical Engineering & Applied Earth Sciences.

Additional formats Print version: Ground improvement techniques and geosynthetics. 9789811305580 (DLC) 2018941217

	Printed edition: 9789811305580
	Printed edition: 9789811305603
	Printed edition: 9789811344503
Series	Lecture Notes in Civil Engineering, 2366-2557; 14
	Lecture Notes in Civil Engineering, 2366-2557; 14

Handbook of advanced radioactive waste conditioning technologies

LCCN	2013427745
Type of material	Book
Main title	Handbook of advanced radioactive waste conditioning technologies / edited by Michael I. Ojovan.
Published/Created	Cambridge, UK; Philadelphia, PA: Woodhead Pub, c2011.
Description	xv, 488 p.: ill.; 24 cm.
ISBN	9781845696269 (print)
	1845696263 (print)
	9780857090959 (online)
	085709095X (online)
LC classification	TD898 .H358 2011
Related names	Ojovan, Michael I.
Summary	Annotation Conditioning technology is essential to the nuclear fuel cycle for the immobilisation and encapsulation of spent nuclear fuels and radioactive wastes, forming the initial engineer barrier required for radioactive waste transportation, storage and disposal. Long term radiological safety is ensured through the performance and durability of conditioned waste, which advanced conditioning technologies and techniques continue to improve upon. This book provides a comprehensive and systematic reference on advanced radioactive waste conditioning technologies. Coverage includes fundamental science and safety criteria, treatment processes and immobilisation technologies, the development of specific types of waste form materials and systems, and performance and safety analysis - Source other than Library of Congress.

Contents Radioactive waste characterization and selection of processing technologies / M.I. Ojovan - pt. 1. Radioactive waste treatment processes and conditioning technologies. Compaction processes and technology for treatment and conditioning of radioactive wastes / M. Garamszeghy - Incineration and plasma processes and technology for treatment and conditioning of radioactive waste / J. Deckers - Application of inorganic cements to the conditioning and immobilisation of radioactive wastes / F. Glasser - Calcination and vitrification processes for conditioning of radioactive wastes / A.S. Aloy - Historical development of glass and ceramic waste forms for high level radioactive wastes / C.M. Jantzen - Decommissioning of nuclear facilities and environmental remediation: generation and management of radioactive and other wastes / M. Laraia - pt. 2. Advanced materials and technologies for the immobilisation of radioactive wastes. Development of geopolymers for nuclear waste immobilisation / E.R. Vance and D.S. Perera - Development of glass matrices for high level radioactive wastes / C.M. Jantzen - Development of ceramic matrices for high level radioactive wastes / H. Kinoshita - Development of waste packages for the disposal of radioactive waste: French experience / G. Ouzounian and R. Poisson - Development and use of metal containers for the disposal of radioactive wastes / I.G. Crossland-- pt. 3. Radioactive waste long-term performance assessment and knowledge management techniques. Failure mechanisms of high level nuclear waste forms in storage and geological disposal conditions / V.V. Rondinella - Development of long-term behavior models for radioactive waste forms / D. Bacon and E. Pierce - Knowledge management for radioactive waste management organisations / P. Gowin ... [et al.].

LC Subjects	Radioactive waste disposal--Handbooks, manuals, etc.
Notes	Includes bibliographical references and index.
Series	Woodhead Publishing series in energy, 2044-9364; no. 12 Woodhead Publishing in energy; no. 12.

High-performance materials from bio-based feedstocks

LCCN	2021052433
Type of material	Book
Main title	High-performance materials from bio-based feedstocks / [edited by] Andrew J. Hunt, Nontipa Supanchaiyamat, Kaewta Jetsrisuparb, Jesper T. Knijnenb.
Edition	First edition.
Published/Produced	Hoboken, NJ: Wiley, 2022.
ISBN	9781119655725 (cloth) (adobe pdf) (epub)
LC classification	TP248.B55 H539 2022
Related names	Hunt, Andrew J., editor. Supanchaiyamat, Nontipa, editor. Jetsrisuparb, Kaewta, editor. Knijnenb, Jesper T., editor.
Summary	"This book presents recent developments in the research, manufacture and application of advanced materials from bio-based feedstocks. Focusing not only on bio-based polymers, but also on inorganic components of biomass and conversion of biomass to advanced materials, the wealth of potential offered by such systems is illuminated. The structural properties of materials are correlated with various real life applications such as use in medicine (as tissue scaffolds or drug delivery), in construction (for greener pavements or buildings), in synthesis (as catalysts or solvents), in energy storage (for batteries or supercapacitors), in agriculture (for slow release fertilizers or remediation), and in food (packaging materials and

nutrient delivery). This book targets many such applications and presents deep insights into the materials production, properties and performance. Topics covered include: *Advanced materials for chemistry: *Biomass to medical devices *Advanced bio-based inorganic materials *Advanced bio-based materials for energy storage applications *From farm to advanced materials"--Provided by publisher.

Contents	High-performance materials from bio-based feedstocks: introduction and structure of the book - Bio-based carbon materials for catalysis - Starbon® - novel template-free mesoporous carbonaceous materials from biomass - synthesis, functionalisation and applications in adsorption and catalysis - Conversion of biowastes into carbon-based electrodes - Bio-based materials in electrochemical applications - Bio-based materials using deep eutectic solvent modifiers - Biopolymer composites for recovery of precious and rare earth metals - Bio-based materials in anti-HIV drug delivery - Chitin- a natural bio-feed stock and its derivatives: chemistry and properties for biomedical applications - Carbohydrate-based materials for biomedical applications - Organic feedstock as biomaterial for tissue engineering - Green synthesis of bio-based metal-organic frameworks - Geopolymers based on biomass ash and bio-based additives for construction industry - The role of bio-based excipients in the formulation of lipophilic nutraceuticals - Bio-derived polymers for packaging - Recent developments in controlled release fertilizers from bio-based materials.
LC Subjects	Biomass chemicals.
Notes	Includes bibliographical references and index.
Additional formats	Online version: High-performance materials from bio-based feedstocks. First edition Hoboken, NJ: Wiley, 2022 9781119655732 (DLC) 2021052434
Series	Wiley series in renewable resources

High-performance materials from bio-based feedstocks

LCCN	2021052434
Type of material	Book
Main title	High-performance materials from bio-based feedstocks / [edited by] Andrew J. Hunt, Nontipa Supanchaiyamat, Kaewta Jetsrisuparb, Jesper T. Knijnenb.
Edition	First edition.
Published/Produced	Hoboken, NJ: Wiley, 2022.
Description	1 online resource
ISBN	9781119655626 (epub) 9781119655732 (adobe pdf) (cloth)
LC classification	TP248.B55
Related names	Hunt, Andrew J., editor. Supanchaiyamat, Nontipa, editor. Jetsrisuparb, Kaewta, editor. Knijnenb, Jesper T., editor.
Summary	"This book presents recent developments in the research, manufacture and application of advanced materials from bio-based feedstocks. Focusing not only on bio-based polymers, but also on inorganic components of biomass and conversion of biomass to advanced materials, the wealth of potential offered by such systems is illuminated. The structural properties of materials are correlated with various real life applications such as use in medicine (as tissue scaffolds or drug delivery), in construction (for greener pavements or buildings), in synthesis (as catalysts or solvents), in energy storage (for batteries or supercapacitors), in agriculture (for slow release fertilizers or remediation), and in food (packaging materials and nutrient delivery). This book targets many such applications and presents deep insights into the materials production, properties and performance. Topics covered include: *Advanced materials for chemistry: *Biomass to medical devices *Advanced bio-based inorganic materials

	*Advanced bio-based materials for energy storage applications *From farm to advanced materials"--Provided by publisher.
Contents	High-performance materials from bio-based feedstocks: introduction and structure of the book - Bio-based carbon materials for catalysis - Starbon® - novel template-free mesoporous carbonaceous materials from biomass - synthesis, functionalisation and applications in adsorption and catalysis - Conversion of biowastes into carbon-based electrodes - Bio-based materials in electrochemical applications - Bio-based materials using deep eutectic solvent modifiers - Biopolymer composites for recovery of precious and rare earth metals - Bio-based materials in anti-HIV drug delivery - Chitin- a natural bio-feed stock and its derivatives: chemistry and properties for biomedical applications - Carbohydrate-based materials for biomedical applications - Organic feedstock as biomaterial for tissue engineering - Green synthesis of bio-based metal-organic frameworks - Geopolymers based on biomass ash and bio-based additives for construction industry - The role of bio-based excipients in the formulation of lipophilic nutraceuticals - Bio-derived polymers for packaging - Recent developments in controlled release fertilizers from bio-based materials.
LC Subjects	Biomass chemicals.
Notes	Includes bibliographical references and index.
Additional formats	Print version: High-performance materials from bio-based feedstocks First edition. Hoboken, NJ: Wiley, 2022 9781119655725 (DLC) 2021052433
Series	Wiley series in renewable resources

Innovations and Technologies in Construction: Special topic volume with invited peer-reviewed papers only

LCCN	2020488187
Type of material	Book

Main title	Innovations and Technologies in Construction: Special topic volume with invited peer-reviewed papers only / edited by Sergey V. Klyuev.
Published/Produced	Switzerland: Trans Tech Publications Ltd, 2021.
Description	174 pages: illustrations; 24 cm.
Links	Papers (PDF) on Scientific.net https://www.scientific. net/MSF.1017 Online eBook on Scientific.net https://www.scientific. net/book/innovations-and-technologies-in-construction/978-3-0357-3825-4/ebook https://doi.org/10.4028/www.scientific.net/MSF.1017 Book description https://www.scientific.net/book/ innovations-and-technologies-in-construction/978-3-0357-3825-4 http://public.eblib.com/choice/PublicFullRecord.aspx?p=6427672 EBSCOhost http://search.ebscohost.com/login.aspx?direct=true&scope=site&db=nlebk&db=nlabk&AN=2711241
ISBN	9783035738254 3035738254
Related names	Klyuev, S.V. editor.
Summary	The important problems of Building Constructions, Building Materials, Industrial and Civil Construction, and Structural Mechanics and Theory of Structures (from ideas and projects to immediate implementation) are raised, and the ways of their solution and optimization are suggested in the reports presented in frameworks of these topics. Many reports were prepared as a benefit of the state support of science in the forms of grants, federal target programs, resolutions of the Government of the Russian Federation, and other programs. Concrete, Cement, Composite Binder, Geopolymers, Mechanical Properties, 3D Concrete Printing, Clay, Asphalt Binder, Dry

	Milling, Soil Stabilisation, Mineral Powder, Road Construction Materials Science, Building Materials, Construction.
Contents	Effect of Mixing Procedure and Chemical Composition on Physical and Mechanical Performance of Geopolymers - Effective Application of 3D Additive Technologies for Creation of Construction Compositions Using Technogenic Raw Materials - Effects of Calcination Temperature and Time on the Physical-Chemical Efficiency of Thermally Activated Clays in Cement Systems - The Method of Creating and Measuring the Printability of Fine-Grained Concrete - The Effect of Composition and Fineness of Mineral Fillers on Structure of Asphalt Binder Microstructural Analysis of Changes in the Morphology of Quartz Raw Materials of Different Genesis at Dry Milling - Analysis of the Effectiveness of the New Generation of Additives in Improving Physical and Mechanical Properties of Highway Subgrades - Glass Micro Balls Based on Glass Domestic Waste for Road Construction - Characterization of Different Brazilian Soils for the Production of Ceramic Artifacts - Comparative Resistance of Concrete with Aggregates and Fillers of Different Composition - Recipes and Properties of Multicomponent Systems ""Aspiration Dust - Mineral Powder - Na2SiO3 Specifics of Structure Formation of Cement Stone with Silica Raw Material - A Substrate of Photocatalytic Composite Material - Effectiveness Evaluation Criteria of Using Fine Powders of Secondary Concrete as a Component of Composite Binder - Keyword Index - Author Index
LC Subjects	Building. Building materials. Construction industry--Management.
Notes	Includes bibliographical references and index.

Series	Materials science forum; Vol. 1017 Materials science forum; v. 1017.

International Congress on Polymers in Concrete (ICPIC 2018): Polymers for Resilient and Sustainable Concrete Infrastructure

LCCN	2019744419
Type of material	Book
Main title	International Congress on Polymers in Concrete (ICPIC 2018): Polymers for Resilient and Sustainable Concrete Infrastructure / edited by Mahmoud M. Reda Taha.
Edition	1st ed. 2018.
Published/Produced	Cham: Springer International Publishing: Imprint: Springer, 2018.
Description	1 online resource (XV, 757 pages 336 illustrations, 236 illustrations in color.) text file PDF
ISBN	9783319781754
Related names	Taha, Mahmoud M. Reda. editor.
Summary	This volume collects the proceedings from the International Congress of Polymers in Concrete 2018 (ICPIC), held under the theme "Polymers for Resilient and Sustainable Concrete Infrastructure." ICPIC 2018 provides an opportunity for researchers and specialists working in the fields of polymers to exchange ideas and follow the latest progress in the use of polymers in concrete infrastructure. It also showcases the use of polymers and polymer concrete in sustainable and resilient development, and provides a platform for local and overseas suppliers, developers, manufacturers and contractors using polymers, polymer concrete and polymer composites in concrete structures to develop new business opportunities and follow the latest developments in the field. The International Congress of Polymers in Concrete is an international forum that has taken place every three years for the last 40 years with the objective of following progress in the field of

polymers and their use in concrete and construction. Following 15 successful congresses held in London (1975), Austin (1978), Koriyama (1981), Darmstadt (1984), Brighton (1987), Shanghai (1990), Moscow (1992), Oostende (1995), Bologna (1998), Honolulu (2001), Berlin (2004), Chuncheon (2007), Funchal (2010), Shanghai (2013) and Singapore (2015), the 16th ICPIC will take place in Washington, DC, from April 29 to May 1st, 2018.

Contents Nano Silica on the Strength of Natural Pozzolan-Based Alkali Activated Concrete - Part IX Strengthening and Restoration Using Polymers - Review of Polymer Coatings Used for Blast Strengthening of Reinforced Concrete and Masonry Structures - Evaluation of Polymer-Modified Restoration Mortars for Corrosion Resistance of Sewage Treatment Structures - Finite Element Modeling of CFRP Strengthened Low Strength Concrete Short Columns - Improvement Works to Existing Column Stumps by Fiber Reinforced Polymer Strengthening System - Silicone Resin Enclosing Method Applied for the Maintenance of Steel Bearings - Bio-based Polyurethane Elastomer for Strengthening Application of Concrete Structures under Dynamic Loadings.

Part I Keynote Papers - 1 Concrete-Polymer Materials: How Far Have We Come and Were Do We Need to Go? - Polymer Concrete for Bridge Preservation - Feasibility Study of the Use of Polymer-Modified Cement Composites as 3D Concrete Printing Material - Experimental Analysis and Micromechanics-Based Prediction of the Elastic and Creep Properties of Polymer-Modified Concrete at Early Ages - Durability and Long-Term Performance of Fiber-Reinforced Polymer as a New Civil Engineering Material - Nano-modified Polymer Concrete - A New

Material for Smart and Resilient Structures - Part II Polymer Materials - Bio-Based Superplasticizers for Cement Based Materials - Screening Encapsulated Polymeric Healing Agents for Carbonation Exposed Self-Healing Concrete, Service Life Extension and Environmental Benefit - Synthesis and Characterization of Superabsorbent Polymer Hydrogels Used as Internal Curing Agents: Impact of Particle Shape on Mortar Compressive Strength - Evaluation of Microencapsulated Corrosion Inhibitors in Reinforced Concrete - The Use of Polymer Additions to Enhance the Performance of Industrial and Residential Decorative Concrete Flooring - The Effect of Glucose on the Properties of Cement Paste - Microstructured Polymers and Their Influences on the Mechanical Properties of PCC - Stability of Latex in Cement Paste: Experimental Study and Theoretical Analysis - Experimental Verification of Use Secondary Raw Materials as Fillers to Epoxy Polymer Concrete - Effects of Anionic Asphalt Emulsion on Early Age Cement Hydration - Polymer Solutions for Protection of Concrete Exposed to Strong Alkaline or Acid Effluent on Industrial Installations - Lightweight Filled Epoxy Resins for Timber Restoration - Effect of Methyl Methacrylate Monomer on Properties of Unsaturated Polyester Resin-Based Polymer Concrete - Analysis of Mechanical Behavior and Durability of Coatings for Use as Flooring in the Petroleum Industry - Evaluation of the Performance of Engineered Cementitious Composites (ECC) Produced from Local Materials - Application of Phase Change Materials (PCM) in Concrete for Thermal Energy Storage - Mortars with Phase Change Materials and Stone Waste to Improve Energy Efficiency in Buildings - Physical and Mechanical Properties of Cement Mortars with Direct Incorporation of Phase Change Material -

Mechanical Performance of Fly Ash Geopolymeric Mortars Containing Phase Change Materials - Part III Polymer Concrete - Are Polymers Still Driving Forces in Concrete Technology? - Mechanical Properties of Polymer Cement - Fiber Reinforced Concrete (PC-FRC) - Comparison Based on Experimental Studies - Environmental Temperature and Humidity Adaptability of Polymer Modified Cement Mortar - Innovative Polymer-Modified Pervious Concrete - Combined Methods to Investigate the Crack-Bridging Ability of Waterproofing Membranes - Polymer Concrete for A Modular Construction System - Investigation of Mechanical Properties and Bond Behaviour by Means of X-Ray CT - Bending and Crack Characteristics of Polymer Lattice-Reinforced Mortar - The Influence of Specimen Shape and Size on The PCC Compressive Strength Values - Long-term Investigation on the Compressive Strength of Polymer Concrete with Fly Ash - Overlays: A Great Use for Polymer Concrete - PC with Superior Ductility using Mixture of Pristine and Functionalized Carbon Nanotubes - Contribution of Concrete-Polymer Composites and Ancient Mortar Technology to Sustainable Construction - Smart Monitoring of Movement and Internal Temperature Changes Within Polymer Modified Concrete Repair Patches - PIC: Does It Have Potential? - A Perspective on 40 Years of Polymers in Concrete History - Development Length of Steel Reinforcement in Polymer Concrete for Bridge Deck Closure - Development of Ultra Rapid-Hardening Epoxy Mortar for Railway Sleepers - Contribution of C-PC to Resilience of Concrete Structures in Seismic Country Japan - Precast Polymer Concrete Panels for Use on Bridges and Tunnels - Part IV Polymer Fiber Concrete - The Effect of Combinations of Treated Polypropylene Fibers on the Energy

Absorption of Fiber Reinforced Shotcrete - Bond Performance of Steel Reinforced Polymer (SRP) Subjected to Environmental Conditioning and Sustained Stress - High-Strength Strain-Hardening Cement-Based Composites (HS-SHCC) Made with Different High-Performance Polymer Fibers - Uniaxial Tensile Creep Behavior of Two Types of Polypropylene Fiber Reinforced Concrete - Dynamic Behavior of Textile Reinforced Polymer Concrete Using Split Hopkinson Pressure Bar - Steel-Fiber Self-Consolidating Rubberized Concrete Subjected to Impact Loading - Effect of Fiber Combinations on the Engineering Properties of High Performance Fiber Reinforced Cement Composites - Application of Fibre Reinforced Polymer Reinforced Concrete for Low Level Radioactive Waste Disposal - Efficiency of Polymer Fibers in Lightweight Plaster - Part V Polymer Concrete with Recycled Waste - Properties of Ceramic Waste Powder Based Geopolymer Concrete - Use of Recycled Polymers in Asphalt Concrete for Infrastructural Applications - Influence of Method of Preparation of PC Mortar with Waste Perlite Powder on Its Rheological Properties - Design and Manufacture of a Sustainable Lightweight Prefabricated Material Based on Gypsum Mortar with Semi-Rigid Polyurethane Foam Waste - Cement Mortars Lightened with Rigid Polyurethane Foam Waste Applied on Site. Suitability and Durability - Latex-Modified Concrete Overlays Using Recycled Waste Paint - Effect of Using Kaolin and Ground Granulated Blast-furnace Slag on Green Concrete Properties - Lightweight Structural Recycled Mortars Fabricated with Polyurethane and Surfactants - Hydration in Mortars Manufactured with Ladle Furnace Slag (LFS) and the Latest-Generation of Polymeric Emulsion Admixtures - Chemical Resistance of Vinyl-ester Concrete with

Waste Mineral Dust Remaining After Preparation of Aggregate for Asphalt Mixture - Part VI Geopolymers - Performance Studies on Self Compacting Geo Polymer Concrete at Ambient Curing Condition - Effect of 3D Printing on Mechanical Properties of Fly Ash Based Inorganic Geopolymer - Optimization of Fly Ash Based Geopolymer Using a Dynamic Approach of the Taguchi Method - Microstructural and Strength Investigation of Geopolymer Concrete with Natural Pozzolan and Micro Silica - Performance of Steel-Fiber-Reinforced High Performance One-Part Geopolymer Concrete - Effect of Different Class C Fly Ashes Composition on the Properties of the Alkali Activated Concrete - Effect of Curing Temperatures on Zero-Cement Alkali-Activated Mortars - Properties of PVA Fiber Reinforced Geopolymer Mortar - Thermal Performance of Fly Ash Geopolymeric Mortars Containing Phase Change Materials - Development of Fiber-Reinforced Slag-Based Geopolymer Concrete Containing Lightweight Aggregates Produced by Granulation of Petrit-T - Applications of Geopolymers in Concrete for Low-Level Radioactive Waste Containers - Part VII Fiber Reinforced Polymers (FRP) - Microstructure and Mechanical Property Behavior of FRP Reinforcement Autopsied from Bridge Structures Subjected to In-situ Exposure - The Influences of Mechanical Load on FRP Composites Subjected to Environmental Corrosion - Finite Element Analysis of RC Beams Strengthened in Shear with NSM FRP Rods - Effect of Sustained Load Level on Long-term Deflections in GFRP and Steel-Reinforced Concrete Beams - Flexural Behavior and Cracks in Concrete Beams Reinforced with GFRP Bars - Flexural Rigidity Evaluation of Seismic Performance of Hollow-Core Composite Bridge Columns - Three-Dimensional Numerical

Analysis of Hollow-Core Composite Building Columns - Pultruded GFRP Reinforcing Bars with Carbon Nanotubes - On Mechanical Characteristics of HFRP Bars with Various Types of Hybridization - Fatigue Behavior Characterization of Superelastic Shape Memory Alloy Fiber-reinforced Polymer Composites - Strength Performance of Concrete Beams Reinforced with BFRP Bars - Part VIII PC with Nanomaterial - A Comparative Study on Colloidal Nano Silica Incorporation in Polymer Modified Cement Mortars - Performance of UHPC and Nano-Modified Polymer Concrete (NMPC) Composite Wall Panels for protective Structures - Mechanical Characterization of Polymer Nanocomposites Reinforced with Graphene Nanoplatelets - Oilwell Cement Modified with Bacterial Nanocellulose - Effect of Incorporating

LC Subjects	Building materials. Polymers.
Other Subjects	Building Materials. Polymer Sciences. Structural Materials.
Additional formats	Print version: International congress on polymers in concrete (ICPIC 2018). 9783319781747 (DLC) 2018936519 Printed edition: 9783030086336 Printed edition: 9783319781747 Printed edition: 9783319781761
Other standard no.	10.1007/978-3-319-78175-4

Lea's chemistry of cement and concrete

LCCN	2021277066
Type of material	Book
Personal name	Lea, F. M. (Frederick Measham), 1900-1984, author.
Uniform title	Chemistry of cement and concrete
Main title	Lea's chemistry of cement and concrete / edited by Peter C. Hewlett, PhD, LLD, BSc, CChem, CSci,

	FRSC, FIMM, FInstConcTech, FConcSoc, Martin Liska, PhD (Cantab), MSc, A.M.I.C.T.
Edition	Fifth edition.
Published/Produced	Oxford [England]; Cambridge, MA: Butterworth-Heinemann, [2019]
Description	xxxvii, 858 pages: illustrations; 29 cm
ISBN	9780081007730
LC classification	TA434 .L43 2019
Portion of title	Chemistry of cement and concrete
Related names	Hewlett, P. C., editor. Liska, Martin, editor.
Contents	The History of Calcareous Cements / Edwin A.R.Trout - Manufacture of Portland Cement / Peter del Strother - Components in Portland Cement Clinker and Their Phase Relationships / Duncan Herfort and Donald E. Macphee - Constitution and Specification of Portland Cement / Arthur Michael Harrisson - Hydration, Setting and Hardening of Portland Cement / James Beaudoin and Ivan Odler - Resistance of Concrete to Destructive Agencies / Sidney Mindess - Physicochemical and Mechanical Properties of Portland Cements / Kevin A. Paine - The Production of Low Energy Cements / Thomas Harrison, Martyn Roderick Jones and David Lawrence - Pozzolanas and Pozzolanic Materials / Michael JohnMcCarthy and Thomas Daniel Dyer - Cements Made From Blastfurnace Slag / Micheline Moranville-Regourd and Siham Kamali-Bernard - Microsilica as an Addition / Robert Lewis and Per Fidjestøla - Calcium Aluminate Cements / Jason H. Ideker, Karen L. Scrivener, Hervé Fryda and BrunoTouzo - Special Cements / Martin Liska, Alexander Wilson and John Bensted - Cement and Concrete Admixtures / Peter Clive Hewlett, Harald Justnes and Rodney M. Edmeades - Concrete Aggregates / Ian Sims, John Lay and James Ferrari - Geopolymers and Other Alkali-Activated Materials / John L. Provis and Jannie S.J. van

	Deventer - The Influence of the Water/Cement Ratio on the Sustainability of Concrete / Pierre-Claude Aïtcin.
LC Subjects	Cement. Concrete--Chemistry.
Notes	Includes bibliographical references and indexes.

New Developments in Materials for Infrastructure Sustainability and the Contemporary Issues in Geo-environmental Engineering: Proceedings of the 5th GeoChina International Conference 2018 - Civil Infrastructures Confronting Severe Weathers and Climate Changes: From Failure to Sustainability, held on July 23 to 25, 2018 in HangZhou, China

LCCN	2019768548
Type of material	Book
Main title	New Developments in Materials for Infrastructure Sustainability and the Contemporary Issues in Geo-environmental Engineering: Proceedings of the 5th GeoChina International Conference 2018 - Civil Infrastructures Confronting Severe Weathers and Climate Changes: From Failure to Sustainability, held on July 23 to 25, 2018 in HangZhou, China / edited by Shanzhi Shu, Liangcai He, Yao Kai.
Edition	1st ed. 2019.
Published/Produced	Cham: Springer International Publishing: Imprint: Springer, 2019.
Description	1 online resource (IX, 272 pages 160 illustrations, 84 illustrations in color.) text file PDF
ISBN	9783319957746
Related names	He, Liangcai, editor. Kai, Yao, editor. Shu, Shanzhi, editor.
Summary	This book contains research papers focusing on recent advances throughout the world in theories and technologies of geotechnical engineering and the relevant disciplines. Topics includes: numerical modeling, earthquake engineering, geomaterial application in soil improvement and geo-

environmental engineering, foundation engineering, and geo-environmental engineering.. Papers were selected from the 5th GeoChina International Conference 2018 - Civil Infrastructures Confronting Severe Weathers and Climate Changes: From Failure to Sustainability, held on July 23 to 25, 2018 in HangZhou, China.

Contents 1. Impact of Initial In-Situ Stress Field On Soil Response During Cavity Expansion Using Discrete Element Simulation - 2. Equivalent FEM meshes from axisymmetric (AXID) to three (3D) dimensions applied to tunnels in clay – 3 .Earthquake-induced deformation of breakwater on liquefiable soil with and without remediation: case study of Iran LNG Port - 4. Effect of the seismic vulnerability of water pipelines on the collapsible soils of the North of Chile - 5. Experimental study on the dynamic response of saturated sandy soil with different clayey particle content and skeleton sand size - 6. Model-scale study on the effect of cyclic loading on pile lateral bearing capacity at different directions - 7. Evaluation of Concrete Bored Piles Behaviour in Saturated Loose and Dense Sand during the Static Load Testing - 8. Experimental Investigations on uplift capacities of single and group of granular anchor piles - 9. Generalized Solutions for Lateral Bearing Behavior of Large Diameter Monopile Foundation for Offshore Wind Turbine Considering Double Additional Moment Effects - 10. Full-scale lateral load tests to determine load-displacement characteristics of driven piles in soft clay - 11. An Experimental Study on Strength Characteristics of Cohesionless Soil under Small Gravity Fields - 12. Reasons for mid-span failure of pile supported bridges in case of subsurface liquefaction - 13. Experimental study on Gas permeability of intact loess under applied load with constant stress ratio paths - 14. Weathered Swelling Mudstone

	Landslide and Mitigation Measures in the Yanji Basin: A Case Study - 15. Soil Improvement by PVD in an Harbor Storage Area - 16. Research of Geopolymer deal with the Strength of Soft Soil and Microstructure Test - 17. Mechanical Properties of geopolymers cured in saline water - 18. Water permeability reduction in THF hydrate-bearing sediments - 19. Concept of a Geotechnical Solution to Address the Issues of Sea Water Intrusion in Ashtamudi Lake, Kerala - 20. Importance of Indoor Environmental Quality Criteria to Occupants of Low Income Housing.
LC Subjects	Engineering geology. Engineering--Geology. Foundations. Geotechnical engineering. Hydraulics. Numerical analysis.
Other Subjects	Geotechnical Engineering & Applied Earth Sciences. Geoengineering, Foundations, Hydraulics. Numerical Analysis.
Additional formats	Print version: New developments in materials for infrastructure sustainability and the contemporary issues in geo-environmental engineering. 9783319957739 (DLC) 2018948641 Printed edition: 9783319957739 Printed edition: 9783319957753
Series	Sustainable Civil Infrastructures, 2366-3405 Sustainable Civil Infrastructures, 2366-3405

Non-traditional cement and concrete: 6th International Conference Non-Traditional Cement and Concrete 2017: selected, peer reviewed papers from the 6th International Conference Non-Traditional Cement and Concrete 2017, June 19-22, 2017, Brno, Czech Republic

LCCN	2021378996
Type of material	Book
Meeting name	International Conference Non-Traditional Cement and Concrete (6th: 2017: Brno, Czech Republic)

Main title	Non-traditional cement and concrete: 6th International Conference Non-Traditional Cement and Concrete 2017: selected, peer reviewed papers from the 6th International Conference Non-Traditional Cement and Concrete 2017, June 19-22, 2017, Brno, Czech Republic / edited by Šárka Nenadálová, Vlastimil Bílek, Zbyněk Keršner, Stanislav Seitl and Hana Šimonová.
Published/Produced	Zurich, Switzerland: Scientific.Net: distributed worldwide by Trans Tech Publications, Ltd, [2018]
Description	ix, 216 pages: illustrations; 24 cm.
ISBN	3035710961 (pbk.) 9783035710960 (pbk.)
LC classification	TA434 .I58 2017
Related names	Nenadálová, Šárka, editor.
Summary	"The selected papers include recent results of scientific research in area of non traditional cements and concretes, especially alkali activated materials, geopolymers, clinker-free binders, durability, high performance concrete, microstructure study, etc."--Preface.
LC Subjects	Cement--Congresses. Concrete--Congresses.
Other Subjects	Cement. Concrete.
Form/Genre	Conference papers and proceedings. Conference papers and proceedings.
Notes	Includes bibliographical references and indexes.
Series	Key engineering materials, 1013-9826; volume 761 Key engineering materials; v. 761.

Properties and application of geopolymers: selected, peer reviewed papers from the 2nd Malaysia-Indonesia Geopolymer Symposium, October 25-27, 2015, Surabaya, Indonesia

LCCN	2021378839
Type of material	Book
Meeting name	Malaysia-Indonesia Geopolymer Symposium (2nd: 2015: Surabaya, Indonesia)

Main title Properties and application of geopolymers: selected, peer reviewed papers from the 2nd Malaysia-Indonesia Geopolymer Symposium, October 25-27, 2015, Surabaya, Indonesia / edited by Januarti Jaya Ekaputri and Antoni.
Published/Produced Pfaffikon, Switzerland: TPP, Trans Tech Publications Ltd, [2016]
Description ix, 214 pages: illustrations; 24 cm.
Links 1850-9999 http://www.ttp.net/
ISBN 3038355445 paperback
LC classification QD196 .M35 2015
Related names Januarti Jaya Ekaputri, editor.
Antoni, editor.
LC Subjects Inorganic polymers--Congresses.
Aggregates (Building materials)--Testing--Congresses.
Binders (Materials)--Testing--Congresses.
Notes Includes bibliographical references and index.
Additional formats Also issued online.
Series Materials science forum, 0255-5476; volume 841

Strategic materials and computational design: a collection of papers presented at the 34th International Conference on Advanced Ceramics and Composites, January 24-29, 2010, Daytona Beach, Florida
LCCN 2011275098
Type of material Book
Meeting name International Conference on Advanced Ceramics and Composites (34th: 2010: Daytona Beach, Fla.)
Main title Strategic materials and computational design: a collection of papers presented at the 34th International Conference on Advanced Ceramics and Composites, January 24-29, 2010, Daytona Beach, Florida / edited by Waltrud M. Kriven, Yanchun Zhou, Miladin Radovic; volume editors, Sanjay Mathur, Tatsuki Ohji.
Published/Created Hoboken, N.J.: Wiley, c2010.
Description xii, 192 p.: ill.; 25 cm.
Links Publisher description http://www.loc.gov/catdir/enhancements/fy1203/2011275098-d.html

	Table of contents only http://www.loc.gov/catdir/enhancements/fy1203/2011275098-t.html
ISBN	9780470921913 0470921919
LC classification	TA455.C43 I558 2010b
Related names	Kriven, Waltraud M. Zhou, Yanchun. Radovic, Miladin. Mathur, Sanjay. Ohji, T. (Tatsuki) American Ceramic Society.
LC Subjects	Ceramic materials--Congresses. Composite materials--Congresses. Inorganic polymers--Congresses. Ceramic materials--Computer simulation--Congresses. Composite materials--Computer simulation--Congresses. Carbides--Congresses. Nitrides--Congresses.
Notes	"Contributions from three Focused Sessions that were part of the 34th International Conference on Advanced Ceramics and Composites (ICACC), in Daytona Beach, FL, January 24-29, 2010 are presented in this volume. The broad range of topics is captured by the Focused Session titles, which are listed as follows: FS1--Geopolymers and other Inorganic Polymers; FS3--Computational Design, Modeling Simulation and Characterization of Ceramics and Composites; and FS4--Nanolaminated Ternary Carbides and Nitrides (MAX Phases)."--Pref. "The American Ceramic Society." Includes bibliographical references and index.
Series	Ceramic engineering and science proceedings; v. 31, issue 10 Ceramic engineering and science proceedings; v. 31, no. 10.

Sustainable Construction and Building Materials: Select Proceedings of ICSCBM 2018

LCCN	2019742977
Type of material	Book
Main title	Sustainable Construction and Building Materials: Select Proceedings of ICSCBM 2018 / edited by Bibhuti Bhusan Das, Narayanan Neithalath.
Edition	1st ed. 2019.
Published/Produced	Singapore: Springer Singapore: Imprint: Springer, 2019.
Description	1 online resource (XV, 906 pages 508 illustrations, 434 illustrations in color.) text file PDF
ISBN	9789811333170
Related names	Das, Bibhuti Bhusan. editor. Neithalath, Narayanan. editor.
Summary	This book presents select proceedings of the International Conference on Sustainable Construction and Building Materials (ICSCBM 2018), and examines a range of durable, energy-efficient, and next-generation construction and building materials produced from industrial wastes and byproducts. The topics covered include alternative, eco-friendly construction and building materials, next-generation concretes, energy efficiency in construction, and sustainability in construction project management. The book also discusses various properties and performance attributes of modern-age concretes including their durability, workability, and carbon footprint. As such, it offers a valuable reference for beginners, researchers, and professionals interested in sustainable construction and allied fields.
Contents	Experimental Study on Self Compacting Concrete with Replacement Material's - Performance of Nano-Sio2 and Nano-Zno2 on Compressive Strength and Microstructure Characteristics of Cement Mortar - Fly Ash Utilization In Lightweight Aggregates For Sustainable

Construction - Strength Behaviour of Masonry Blocks Produced Using Green Concrete - Interference of Two Shallow Square Footings on Geogrid Reinforced Crusher Dust - Use of Foundry Sand As Partial Replacement of Natural Fine Aggregate For The Production of Concrete - Construction Blocks From C&D Debris Using The Innovative Co2 Sequestration Technique - Durability Studies on Alkali Activated Fly Ash and Ggbs Based Geopolymer Mortars - A Comparative Study on Rcc Structures (Frame, Infill, Bracings, Wire Frame and Shear Wall) - Comparative Study on Influence of Lead Rubber Bearing on Rc Structures with Flat Slab and Conventional Slab System Under Seismic Loading - Corrosion Inhibitors Behaviour on reinforced concrete - A Review - A Review of The Mechanical Behavior of Substitution Materials In Self-Healing Concrete - Effect of Silica Fume on The Properties of Fly Ash Geopolymer Concrete - Synthesis of Geo-Polymer Coarse Aggregates Using Class-F Fly Ash and Studies on Its Physical Properties - Optimization of Resources by Real Time Correlation Study For Maximizing The Productivity - Study on Mechanical Properties of Cement Concrete For Partial Replacement of Coarse Aggregate by Shredded Plastic and Cement by Fly Ash and Meta-Kaolin - High Performance Concrete (Hpc) - An Innovative Cement Concrete Mix Design To Increase The Life Span of Structures - Study on Mechanical Properties and Leaching of Heavy Metals In The Artificial Produced Fly Ash Aggregates - Feasibility of Producing Class F Fly Ash Geopolymer Mortar with Alkaline Water Containing Sodium Carbonate (Na2co3) - Analytical Study on Disintegration of Concrete - An Effect of Naoh Molarity on Flyash - Metakaolin Based Self Compacting Geopolymer Concrete - A Review on Performance of

Geopolymer In Concrete - An Experimental Study on Workability and Strength Characteristics of M40 Grade Concrete by Partial Replacement of Cement with Nano Tio2 - Studies on Fresh and Hardened Properties of Sugarcane Bagasse Ash Blended Self Compacting Concrete Mixes - Experimental Study on Improvement of Bearing Capacity Using Geosynthetic Stone Column - Analysis and Modeling of Labor Productivity Using System Dynamics Approach In Construction Projects - Implication of Concrete with Chemical Admixture Cured In Low Temperature on Strength, Chloride Permeability and Microstructure - Experimental Investigation on The Properties of Pervious Concrete Over Fibre Reinforced Pervious Concrete - Role of Silica Fume In Producing High Strength Self-Compacting Concrete - Problem Analysis and Geotechnical Study At Sengulam Augmentation Scheme - Graphene In The Domain of Construction: A Review of Applications and Prospects - Experimental Analysis on Partial Replacement of Fine Aggregate by Granite Dust In Concrete - Experimental Investigation of High Strength Self-Compacting Fibre Reinforced Concrete - Effect of Curing Conditions on Mechanical Properties of Reactive Powder Concrete with Different Dosage of Quartz Powder - The Effects of Ggbfs on Strength Properties of Geopolymer Concrete Cured At Ambient Temperature - Experimental and Finite Element Analysis of 80mpa Two-Span High-Performance Concrete Beam Under Flexure - Techniques For Preparation and Dispersion of Nano- Sio2 In Cementitious System - A Review - Life Cycle Costing For The Analysis of Cost-Effectiveness of Alternative Concretes and Masonry Blocks - Study on Development of Strength Properties of Bio-Concrete - Rice Husk Ash (Rha) - The Future of Concrete - Effect of Nano-Silica and Ggbs on The

Strength Properties of Fly Ash-Based Geopolymers - Compressive Strength Prediction of High Strength Concrete Using Regression and Ann Models - Prediction of Compressive Strength of High Volume Fly Ash Concrete Using Artificial Neural Network - Experimental Investigation on Compressive Strength of Ld Slag Aggregate Concrete - Cost Reduction Techniques on Mep Projects - Mineralogical Study of Concretes Prepared Using Carbonated Fly Ash As Part Replacement of Cement - Effect of Supplementary Cementitious Materials on Mechanical Properties and Thermal Conductivity of Concretes and Masonry Blocks - Mechanical Properties of Pavement Quality Concrete Produced with Reclaimed Asphalt Pavement Aggregates - Effect of Partial Replacement of Coarse Aggregates with E-Waste on Strength Properties of Concrete - Experimental Investigation on Utilization of Waste Shredded Rubber Tire As A Replacement To Fine Aggregate In Concrete - Strength Characteristics of Laterized Mortars Using Processed Laterite - Sustainable Building Management by Using Alternative Materials and Techniques - Effect of Silica Fume on Fly-Ash Based Geopolymer Mortar with Recycled Aggregates - Mechanical Properties of Fibre Reinforced Concrete with Bottle Crown Caps As Fibres - Performance of Deep Excavation For An Underground Metro Station Constructed by Top-Down Method - A Case Study - Experimental Investigation on The Strength of Concrete by Partial Replacement of Fine Aggregates by Low Density Shredded Polyethylene - Experimental Study on Performance of M30 Grade Concrete by Partial Replacement with Fly-Ash and Granite Powder - Combined Effect of Marine Environment and Ph on The Impedance of Reinforced Concrete Studied by Electrochemical Impedance Spectroscopy - Study on Effect of Sodium

Hydroxide Concentration on Geopolymer Mortar - Early Cost Estimation of Highway Projects In India Using Artificial Neural Network - Review Paper on Utilization Potential of Rice Husk Ash As Supplementary Cementitious Material - Replacement of Fine Aggregates by Recycled Construction and Demolition Waste In Pavement Quality Concrete - Fresh and Hardened Properties of Self-Consolidating Concrete Incorporating Alumina Silicates - Effect of Various Additives on The Properties of Fly Ash Based Geopolymer Mortar - Influence of Metakaolin and Red-Mud Blended Cement on Reinforcement Corrosion In Presence of Chloride and Sulfate Ions - Durability Studies of Polypropylene Fibre Reinforced Concrete - Durability Studies of Steel Fibre Reinforced Concrete - Durability Studies on Glass Fibre Reinforced Concrete - Comparative Study and Laboratory Investigation of Soil Stabilized Using Terrasil and Zycobond - Partial Replacement of Steel Slag Aggregates In Concrete As Fine Aggregates (Induction Blast Furnace Slag) - Experimental and Numerical Studies on The Behaviour of Broad Gauge Railway Sleepers In Stat[not][not][not]ic Bending Condition - Methods To Monitor Resources and Logistic Planning At Project Sites - Can Geopolymer Concrete Replace The Conventional Concrete? - State of The Art - Ambient Cured Geopolymer Concrete Products - Shrinkage Behaviour of High Strength Concrete Using Recycled Concrete Aggregate - Influence of Nano-Silica on Characteristics of Cement Mortar and Concrete - Durability Performance of Structural Light Weight Concrete - Mechanical Properties of Fiber Reinforced Concrete Using Coal-Bottom Ash As Replacement of Fine Aggregate - An Experimental Study on Mechanical Properties of Ultra-High Performance Fiber Reinforced Concrete (Uhpfrc) - Strength

	Improvement of Cement Mortar In Addition of Ureolytic Microorganism - An Experimental Study on Partial Replacement of Fine Aggregate by Vermiculate and Cement by Marble Powder.
LC Subjects	Building materials. Building. Buildings--Design and construction. Building--Superintendence. Construction industry--Management. Construction. Engineering, Architectural. Sustainable architecture.
Other Subjects	Building Materials. Building Construction and Design. Construction Management. Sustainable Architecture/Green Buildings.
Additional formats	Print version: Sustainable construction and building materials. 9789811333163 (DLC) 2018961224 Printed edition: 9789811333163 Printed edition: 9789811333187
Series	Lecture Notes in Civil Engineering, 2366-2557; 25 Lecture Notes in Civil Engineering, 2366-2557; 25

Sustainable construction materials: municipal incinerated bottom ash

LCCN	2018304269
Type of material	Book
Personal name	Dhir, Ravindra K., author.
Main title	Sustainable construction materials: municipal incinerated bottom ash / Ravindra K. Dhir, OBE, Jorge de Brito, Ciarán J. Lynn, Rui V. Silva.
Published/Created	Duxford, U.K.: Woodhead Publishing, an imprint of Elsevier, [2018]
Description	xiv, 443 pages: illustrations (some color); 24 cm.
ISBN	9780081009970 (print) 0081009976 (print) 9780081009963 (online) 0081009968 (online)
LC classification	TA403.6 .D445 2018

Portion of title	Municipal incinerated bottom ash
Related names	Brito, Jorge de, author. Lynn, Ciarán J., author. Silva, Rui V., author.
Contents	Machine generated contents note: 1. Introduction - Synopsis - 1.1. Background - 1.2. Sustainable Construction Materials - 1.3. Municipal Incinerated Bottom Ash - 1.4. Layout and Contents - References - 2. Methodology - Synopsis - 2.1. Introduction - 2.2. Literature Search and Appraisal - 2.3. Building the Data Matrix - 2.4. Analysis, Evaluation and Modelling of Data - 2.5. Dissemination - 2.6. Conclusions - References - 3. Municipal Solid Waste Composition, Incineration, Processing and Management of Bottom Ashes - Synopsis - 3.1. Introduction - 3.2. Composition of Municipal Solid Waste - 3.3. Municipal Solid Waste Incineration - 3.4. Treatment of Municipal Incinerated Bottom Ash - 3.5. Municipal Incinerated Bottom Ash Management - 3.6. Conclusions - References - 4. Municipal Incinerated Bottom Ash Characteristics - Synopsis - 4.1. Introduction - 4.2. Physical Properties - 4.3. Chemical Properties - 4.4. Engineering Properties - 4.5. Conclusions - References - 5. Concrete-Related Applications - Synopsis - 5.1. Introduction - 5.2. Use as an Aggregate Component - 5.3. Use as a Cement Component - 5.4. Environmental Assessment - 5.5. Case Studies - 5.6. Conclusions - References - 6. Geotechnics and Road Pavements - Synopsis - 6.1. Introduction - 6.2. Geotechnical Properties and Unbound Applications - 6.3. Hydraulically Bound Applications - 6.4. Bituminous-Bound Applications - 6.5. Environmental Assessment - 6.6. Case Studies - 6.7. Conclusions - References - 7. Alternative Applications - Synopsis - 7.1. Introduction - 7.2. Ceramics - 7.3. Agriculture - 7.4. Adsorbent Materials - 7.5. Geopolymers - 7.6. Anaerobic

	Digestion and Landfill Gas Production - 7.7. Insulation Material - 7.8. Environmental Assessment - 7.9. Case Studies - 7.10. Conclusions - References - 8. Environmental Assessment - Synopsis - 8.1. Introduction - 8.2. Leaching Properties of Municipal Incinerated Bottom Ash - 8.3. Artificial Aggregates - 8.4. Cementitious Composites - 8.5. Road Construction - 8.6. Ceramics - 8.7. Conclusions - References - 9. Case Studies and Standards - Synopsis - 9.1. Introduction - 9.2. Incineration of Municipal Solid Waste - 9.3. Municipal Incinerated Bottom Ash Management - 9.4. Municipal Incinerated Bottom Ash Processing and Storage - 9.5. Aggregate Manufacturing - 9.6. Raw Feed in Cement Clinker Production - 9.7. Concrete and Mortar Production - 9.8. Geotechnical Applications - 9.9. Road Pavements - 9.10. Ceramics - 9.11. Landfill - 9.12. Standards and Specifications - 9.13. Conclusions - References - 10. Epilogue.
LC Subjects	Waste products as building materials. Sustainable construction. Municipal solid waste incinerator residues.
Other Subjects	Municipal solid waste incinerator residues. Sustainable construction. Waste products as building materials.
Notes	Includes bibliographical references and index.
Series	Woodhead Publishing series in civil and structural engineering Woodhead Publishing series in civil and structural engineering.

Index

D

E

F

G

S

X

Z